Jan Hixson
7/92

D0172126

WHY PRESERVE
NATURAL VARIETY?

STUDIES IN MORAL, POLITICAL,
AND LEGAL PHILOSOPHY

General Editor: Marshall Cohen

Written under the auspices of the
Center for Philosophy and Public Policy,
University of Maryland

WHY PRESERVE
NATURAL VARIETY?

Bryan G. Norton

PRINCETON UNIVERSITY PRESS
PRINCETON, NEW JERSEY

Copyright © 1987 by Princeton University Press
Published by Princeton University Press
41 William Street
Princeton, New Jersey 08540
In the United Kingdom:
Princeton University Press, Guildford, Surrey

All Rights Reserved

ISBN 0-691-07762-2

Publication of this book has been aided by the
Whitney Darrow Fund of Princeton University Press

This book has been composed in Linotron Sabon

Clothbound editions of Princeton University Press books
are printed on acid-free paper, and binding materials
are chosen for strength and durability. Paperbacks,
although satisfactory for personal collections,
are not usually suitable for library rebinding

Printed in the United States of America
by Princeton University Press
Princeton, New Jersey

For my mother and
in memory of my father

CONTENTS

PREFACE

What is the endangered species problem? When I was invited by the Center for Philosophy and Public Policy to conduct a study of the objectives to be pursued by a national policy to preserve species, I believed that I at least understood the nature of the issue. As for most interested observers the problem was exemplified for me by the controversial and well-publicized situations in which the Endangered Species Act of 1973 had been invoked.

I had used the controversy over the snail darter, known in the legal system as *TVA* v. *Hill*, as a case study in my classes on environmental ethics. My students and I had followed with interest the David and Goliath story that pitted the snail darter, a three-inch member of the perch family, against the Tennessee Valley Authority and its powerful political backers. When the snail darter was discovered, construction was well underway on the $119 millon project, touted as a provider of hydroelectric power, a recreationally attractive lake, and flood abatement.

The Sixth Circuit Court of Appeals blocked completion of the project because it was to flood the shallow waters of a stretch of the Little Tennessee River, destroying the only known habitat of the snail darter. The U.S. Supreme Court upheld this ruling, noting that Section 7 of the Endangered Species Act "admitted of no exception," requiring all federal agencies and departments to "insure that actions authorized, funded, or carried out by them do not jeopardize the continued existence of (any) endangered species. . . ." After Congress weakened the act, creating a high-level committee to rule on exceptions to these general provisions, the underdog snail darters won another round: the "God Committee" concluded unanimously that, far from overriding the value of a species, the Tellico dam had been a worthless pork barrel project from the beginning.

But the celebration was premature. Late-night legislative shenanigans revived Goliath; and, in a denouement that further confused the champions of the underdog, another population of snail darters was discovered in a neighboring branch of the Tennessee River system.

Meanwhile, a legal battle was heating up concerning another large dam proposed to create hydroelectric power on the St. John River in Maine. A botanist had found eight hundred specimens of a small plant, discovered by Kate Furbish early in this century and presumed extinct since, living on the river's bank. Again, the Endangered Species Act appealed to environmentalists fighting this project as the slingshot by which the tiny Furbish lousewort might destroy a champion of the Philistines.

The publicity surrounding these and related cases that tested the limits of application of the act created a serious distortion in the public perception. These controversial cases encouraged a narrow conception of the purposes of an endangered species policy. In the public context just described, it appeared that the Endangered Species Act addressed the question of what to do when a project designed to increase human resources was sited on or near the habitat of a constitutively rare species with a very limited range. The general application of the act required a process for identifying and listing endangered species. This process further encouraged an already existing tendency toward concern for individual species that are naturally rare or that have suffered population declines to the brink of extinction.

Because the endangered species problem gained widespread public attention as existing legislation was used to protect marginal species, a distorted picture of the problem emerged. But at the same time that these celebrated legal battles raged concerning constitutively rare species that appear relatively insignificant ecologically, the enforcement of the act was contributing with much less fanfare to the recovery of the American alligator and the brown pelican, once-numerous species that were ecologically important. While these species never reached the brink of extinction, their wide ranges and large populations had been severely reduced by human activities. Their declining populations thus provided more representative examples of the crisis to which the Endangered Species Act must respond.

I have come to realize, after several years of intense work on endangered species policy, that the endangered species issue is not one, but a cluster of related problems. Viewed most narrowly, it is the question of what a relatively wealthy and fast-growing society should do when it discovers that the population of some species is so depleted that its continued existence is immediately threatened. In this view individual species are identified and a recovery program is proposed. Restrictions on "takings" are enforced and areas of "critical habitat" may be designated and protected. This narrow view of the problem was encouraged because citizens' awareness had developed after an act designed to list and protect every species, taken singly, had been implemented.

Other, more general, questions demand answers. What should be done to ensure that species which still have relatively healthy reproductive populations will not decline toward a minimum threshold where they will require individual attention? What should be done to reverse the general trend toward biological simplicity so evident in North America and, especially, in tropical, developing countries? The endangered species problem can be viewed as a symptom of the pervasive tendency to convert more and more relatively natural ecosystems to intense human use. A thorough examination of endangered species policy must inquire whether this trend can and should continue indefinitely.

In accepting funding from the Ethics and Values in Science and Technology Program of the National Science Foundation for this project, we agreed to ask and propose answers to two questions: (1) What reasons can be given for a policy of perserving species? and (2) Given the most reasonable answer to (1), what should be done if financial and personnel resources are insufficient to protect all species?

These questions are addressed here in the broadest context, the context in which identified, severely endangered species represent symptoms of the more general problems mentioned in the last paragraph. As a result I have interpreted question (1) as requesting a rationale for a general policy protecting biological di-

PREFACE

versity. Why should a society be concerned to limit the destruction of the varied natural ecosystems and associations that provide the habitat for wild species? In this broad view of biological diversity, a species existing in varied habitats presents more diversity than does a species confined to a single "critical habitat." A single species existing in different associations faces varied competitive and selectional regimens. If the concern is with species diversity over the long run, these varied habitats represent alternative ecological and evolutionary trajectories, and this variety is essential to encourage genetic diversity in populations and, in the long run, speciation.

Similarly, I have not merely posed question (2) as one of interspecific priorities for preserving species. This narrow formulation of the priorities question is encouraged if we view the Endangered Species Act as a response to the critical endangerment of identifiable, individual species. But I have just argued that this interpretation of the issue addresses only one aspect of a much more pervasive problem. The priorities question is best posed: How ought insufficient funds and efforts be spent to meet threats to biological diversity? Besides being virtually impossible to apply intelligently, proposals to favor some categories over others fail to address the endangered species problem in its full magnitude. In this broad version, the problem of priorites becomes one of obtaining the best return on public resources. And, given the entire range of concerns that can be designated "the endangered species problem," a systematic effort to preserve habitats will maximize the return on investment of those resources. If this argument is persuasive, the Endangered Species Act, characterized as it is by a predominantly species-by-species approach, may require reexamination. While it is essential to have such an agency to protect individual species in crisis situations, the cases of the snail darter, the Furbish lousewort, and other severely endangered species represent only symptoms of far more pervasive problems. If the society is to address the entire range of concerns regarding biological diversity, it must reexamine its entire style of life and the values that drive it.

Many individuals and institutions contributed to this project. From the beginning I benefited from the knowledge and good judgment of Henry Shue, Director of the Center for Philosophy and Public Policy, during the completion of the project. I began by discussing the project with numerous helpful people from the Washington policy community and from the University of Maryland; I thank them for their time and good advice. Drawing on their suggestions, we assembled a working group, and I thank its members for their contributions to the project. Besides writing chapters for the book, *The Preservation of Species*, they provided me with an interdisciplinary education of incalculable value.

I gratefully acknowledge the support of the National Science Foundation and the National Endowment for the Humanities for the funding through the EVIST program that made the project possible (grant IPS-80-24358). I also thank the program coordinators, Rachelle Hollander and Eric Juengst for their help.

Claudia Mills greatly improved my very rough, rough drafts; readers, as well as I, owe her gratitude. I was assisted in the project by Paula Henry, Michael Heller, Dan Mason, and Merissa Lovett. The support staff of the Center for Philosophy and Public Policy, especially Carroll Linkins and Louise Collins, prepared the manuscript and maintained good cheer against all odds; Shirley Schultz helped me through the final stages of manuscript preparation after my return to New College, and provided moral support. Because this book was a product of a much larger project, its completion owes much to all those who have worked on that project.

I thank, also, several readers who provided helpful suggestions on the manuscript, including Alan Randall, Holmes Rolston, III, Paul Taylor, Mary Anne Warren, and David F. Norton.

WHY PRESERVE
NATURAL VARIETY?

O N E

A RATIONALE FOR PRESERVING
SPECIES: AN APOLOGY AND
A TAXONOMY

1.1 A Rationale: Why

The passage of several endangered species acts, culminating in the comprehensive Endangered Species Act of 1973, represents a series of landmarks for the environmental movement.[1] By these steps the legislative branch of the United States government instituted a policy designed to protect all species from extinction. The passage of the acts resulted from a significant groundswell of public opinion: over a period of a decade it had become fashionable, even *de rigueur*, for politicians to advocate environmental protection. Support for the endangered species acts became a popular means to establish credentials as an environmentalist. The eventual result was a very strong piece of legislation—strong especially in the sense of being highly general in application.[2] This legislation, therefore, has taken on special significance as a "flagship" of the environmental movement.

But at the same time the strength and generality of the act concealed extraordinary diversity in the goals, interests, and viewpoints of those who supported it. Because there was unity in the desire for strong legislation, little attempt was made to discuss and resolve differences in underlying motives and goals. Section

[1] For a history of federal endangered species legislation, see "Comment: Endangered Species Protection: A History of Congressional Action," *Environmental Affairs* 4 (1975): 255.

[2] See Mark Sagoff, "On the Preservation of Species," *Columbia Journal of Environmental Law* 7 (1980): 33-34.

3

2.(a) of the Endangered Species Act of 1973 contains an inclusive list of reasons for preserving nonhuman species.[3] That they are listed in *alphabetical* order is symptomatic of the strategy behind the act: Given agreement on the goal of species preservation, why risk controversy by taking a stand on why, exactly, species should be saved? Since the act protected all species, there was no need to quibble about the reasons for protection.[4] Any supporter could consider his favorite reason to be first and foremost. The acts thus became both a powerful policy statement capable of affecting important social and economic choices and a symbol of a diverse but vocal popular movement.

Two developments have, however, undermined this happy consensus since the institution of the acts. First, in a small number of well-publicized cases critics of large developmental projects were able to use the acts to delay or even halt plans for construction.[5] At this point the generality of the act was called into question. Those who had supported or failed to fight the act in the beginning came to realize that the benefits of a general policy of species preservation have attendant economic costs. The argument that species preservation would always have positive economic effects was therefore questioned. Various attempts to circumvent the act resulted in a landmark Supreme Court case in which the court upheld the generality of the act, arguing:

> One would be hard pressed to find a statutory provision whose terms were any plainer than those in Section 7 of the Endangered Species Act. Its very words affirmatively command all federal agencies "to *insure* that actions *authorized, funded, or carried out* by them do not *jeopardize* the continued existence of an endangered species or *result* in the destruction of modification of habitat of such species. . . ."

16 U.S.C.A. Sec. 1536. [Emphasis in original.] This language

[3] Endangered Species Act of 1973, 16 U.S.C. sec. 1531-1543 (1976; Supplement 1, 1977; Supplement 2, 1978; Supplement 3, 1979).

[4] Protection was not initially equal, however, as penalties for the taking of endangered plants were not included until the 1982 reauthorization.

[5] For a brief history of these controversies, see Phillip Shabecoff, "New Battles over Endangered Species: Birds and Fish vs. Highways and Dams," *New York Times Magazine*, June 4, 1978, pp. 37-44.

admits of no exception. Nonetheless, petitioner urges, as do the dissenters, that the Act cannot reasonably be interpreted as applying to a federal project which was well under way when Congress passed the Endangered Species Act of 1973. To sustain that position, however, we would be forced to ignore the ordinary meaning of plain language.[6]

The effect of this decision was to place the generality of the Endangered Species Act of 1973 back on the congressional agenda, and the result was the passage of an amendment instituting a procedure for considering exceptions to the act. This raised crucial questions about the rationale for the act. As long as the act had perfectly general application, questions of why preservation is important could lie dormant. When exceptions became possible, the "spirit," not just the letter, of the act came under scrutiny, and it became necessary to ask which applications of the hitherto general rule that all species must be protected are most important. This question can be answered only by clarifying the most important reasons for preserving species.

Second, it became clear that simply to decree that all species are to be saved does not automatically ensure that they will be. Implementation of the act necessarily fell short of intent both because legislative appropriations were not adequate to the task set and because the responsible agencies failed to take adequate initiatives to protect particular species. Thus a de facto priority issue arose. Charged with a general mission to save all species, but handicapped by limited budgets and limited scientific and administrative energies, agencies faced inevitable priority choices. The only question was whether these priority decisions would be made according to a rationally discussed and consensually adopted policy or whether they would be left to whim or chance.

Thus, although the Endangered Species Act came into existence with no clearly articulated rationale, and although the need for one was initially avoided because of the generality of the act, day-to-day implementation in a complex political and bureaucratic milieu has progressively demanded decisions regarding its most

[6] *Tennessee Valley Authority* v. *Hill*, 437 U.S. 153 (1978).

important goals. But decisions on when exceptions to the act should be granted and decisions about intraagency priorities can be made on a rational basis only if the various rationales for preserving species are placed in some coherent relationship to one another.

This book, therefore, has three interrelated purposes. First, various reasons for preserving species will be examined and judged in relation to one another. Thus the first task of the book is to develop the most coherent, complete, and internally consistent rationale for preserving species. Second, it will be asked whether the Endangered Species Act, if enforced as written and amended, represents an adequate effort to pursue the values central to this coherent rationale. Third, the thorny issue of priorities of implementation will be addressed in the context of this systematic rationale. If there is to be hope that the explicit or de facto decisions made in implementing the act are rationally justifiable, then they must be made against the backdrop of a rational understanding of the goals of the act. The goals of the act can be understood only if the various rationales for it are evaluated individually and related to each other in a coherent package. Priority decisions consonant with such an examined rationale can be considered rational ones.

1.2 A Taxonomy of Rationales

Public policy discussion, formation, and implementation must rely on the best scientific data; but data, viewed in isolation, imply no goals and objectives. So every policy recommendation includes, explicitly or implicitly, a value premise or premises. And, while it may be possible to classify rationales for preserving species in other ways, a taxonomy based upon value premises can be complete as well as illuminating. Value concepts will thus provide the basis for my taxonomy of rationales for preserving species.

Here I break no new ground; it has become commonplace to distinguish arguments for environmental preservation as "prudential" or "aesthetic," and so on. Using more general value categories, some authors also distinguish "ethical" from "economic" reasons for preservation.

Value premises and the axiological systems supporting them can be categorized further as "anthropocentric" or "nonanthropocentric" depending on whether all value countenanced by them derives ultimately from human values. Value systems normally incorporate a distinction between intrinsic value, the worth some objects have in their own right, independent of their value to any other object, and instrumental value, the usefulness objects have in fulfilling other ends.[7] The anthropocentrist/nonanthropocentrist debate in species preservation reduces to the question of whether any species other than *Homo sapiens* possesses intrinsic value. Anthropocentrists recognize in other species only value instrumental to human ends, while nonanthropocentrists insist that some nonhuman species or specimens of them have value in their own right.

Anthropocentric reasons for preserving species are commonly designated as utilitarian: from an exclusively human point of view, a species has utilitarian value if it is instrumental to human happiness. I will not follow this practice, however, because the concept of utilitarianism is too vague to be useful in the current context. Utilitarianism has come to refer to a cluster of positions sharing little more than a hedonistic theory of value combined with a consequentialist method of evaluating courses of action. In general discussions of moral theory it is often helpful to have available a generic term to distinguish this broad cluster of positions from another group of theories, "deontological" theories

---

[7] Intrinsic value theories can claim independence in two different senses. The value of an individual or species could be (a) independent of any contribution to human values or (b) independent of any human intentional act of valuing them. Following Callicott, I shall use the inclusive sense: a theory will be considered nonanthropocentric if the value in question is independent of human values (even if the occurrence of such value is dependent upon the existence and experience of values). See J. Baird Callicott, "Non-anthropocentric Value Theory and Environmental Ethics," *American Philosophical Quarterly* 21 (1984): 305-306. Paul Taylor prefers the less inclusive categorization, insisting that a theory is truly nonanthropocentric if the value in question (he refers to it as "inherent worth") is independent in both senses (personal communications, September 6 and December 9, 1985). See Section 8.2.

characterized by their emphasis on the relations between value, obligations, and individual rights. When the discussion focuses on obligations transcending species boundaries, however, utilitarianism is not a useful category.

Utilitarianism posits obligations to promote the greatest happiness for the greatest number. In the present case this leaves at least three important questions unanswered. First, whose happiness? It is usually assumed that the answer is obvious—the happiness of all human beings. But where obligations extending beyond humanity are the very point at issue, this answer surely begs the question.

Second, what exactly is to be promoted when one promotes "happiness"? Pleasure? Pain-avoidance? Interests? As specified by whom? Should the choices of the more enlightened count more heavily? Are intellectual pleasures inherently more valuable than sensual ones? Again, in an interspecific context it seems prejudicial to assume answers to these questions because the capacities to enjoy various types of pleasure and happiness are differentially distributed across species. To emphasize some of these capacities over others would prejudice the case in favor of some species over others.

Third, what is the relation between happiness and preference-fulfillment? If happiness can be equated with the fulfillment of individual desires, then perhaps disparate values can be reduced to a single scale by allowing the intensity of the desires fulfilled to stand as a measure of happiness. But other utilitarians argue that preference satisfaction does not necessarily promote happiness: persons may be even less happy once their misguided preferences are fulfilled. Again, since consumptive attitudes toward other species are often argued to reflect misguided preferences, it would be prejudicial to assume that other species' value can be reduced to their value in fulfilling preferences as given.

Thus, as more questions are therefore raised than answered when arguments for preserving species are classified in utilitarian terms, I will abandon that categorization. Dismissal, however, does not obviate the search for other ways of classifying preservationists' reasons for protecting species. I will introduce a new set of categories.

I begin by defining a "merely felt preference" as any desire or need of an individual that can be sated, at least temporarily, by some specifiable experience of that individual.[8] By contrast, a "considered felt preference" is any desire or need that an individual would express or otherwise exhibit after careful deliberation.[9] Careful deliberation is taken to include a judgment that the desire or need is consistent with a rationally adopted world view, which in turn includes a set of fully defended scientific theories, an attendant metaphysical framework interpreting those theories, and a set of rationally developed, fully defended aesthetic and moral ideals.

Felt preferences are normally taken as givens, as the starting points for determining the best course of action to fulfill those preferences. But they are hardly fixed. Their immunity from criticism in some contexts can give way to a critical attitude in others.

Considered preferences are hypothetical desires or needs. They are desires or needs an individual *would have* if certain very stringent, perhaps even practically impossible, conditions were fulfilled. Insofar as they emerge at all, they do so through a long and arduous process of education and self-reflection. If this process were ever completed, however, it would then be possible to take the resulting world view as a given and submit felt preferences to a test of consistency and coherence with it. It is doubtful whether anyone has ever completed the process of rationally supporting a world view defined this broadly. One can, however, at least vaguely imagine such a complete examination.

References to considered preferences remain valuable in spite of their hypothetical nature because one can show that a particular felt preference should be rejected, upon consideration, in favor of another preference if one can show that it is inconsistent

---

[8] This and related definitions have been presented in my "Environmental Ethics and Weak Anthropocentrism," *Environmental Ethics* 6 (1984): 131-148.

[9] Sometimes I will refer to these, more briefly, as "considered preferences." It should be remembered, however, that considered preferences are a subset of felt preferences, the subset that has survived a process of careful deliberation.

with what appears to be the most rationally defensible theory or ideal regarding that preference. For example, a felt preference for gluttony could be shown not to be a considered preference of an individual who is convinced that food supplies are limited, that food not consumed by one person is available to others, that some persons are starving for lack of food, and that human beings have an obligation to provide aid to the starving.

Felt and considered preferences are preferences for certain experiences, such as the experience of eating food or hiking in a wilderness area. One can evaluate the objects of such preferences by saying that an object has "demand value" if it can provide satisfaction for some felt preference, either before or after it has been examined. A good or service has demand value corresponding in magnitude to the intensity of a felt preference that it can potentially satisfy.

I say that an object has "transformative value," as opposed to demand value, if it provides an occasion for examining or altering a felt preference rather than simply satisfying it. Suppose a teenager who only enjoys rock music would gladly pay thirty dollars (if she had it) to attend an upcoming concert featuring her favorite rock star. A ticket to the concert then has a demand value of thirty dollars for her. When the teenager's grandparents give her an envelope containing a birthday card and a ticket, she is at first overjoyed and then crushed to find the ticket affords admission to a performance by a fine symphony orchestra. She tries to give the ticket away (demonstrating that it has no demand value to her), but her parents insist that, to avoid offending her grandparents, she must attend the symphony. Grudgingly she attends only to discover that classical music, when heard in live performance, is exciting and enjoyable. Without renouncing her adulation for rock stars, she now begins to buy classical records as well, attends future symphony concerts whenever possible, and ultimately derives a lifetime of pleasure and fulfillment that might have been denied her had her actual preferences been fulfilled. The ticket that at first had no demand value therefore has significant transformative value, and similar opportunities will, at a later time, have demand value.

Transformative value can be judged either positively or negatively. For example, when parents complain their child is "hanging out with a bad crowd," they attribute negative transformative value to the activity of loitering with peers of questionable character. Logically speaking, attributes of transformative value presuppose that some preferences and values are preferable to others. They are, therefore, relative to a more basic value judgment. In the music example I have attributed to the grandparents only the view that the subsequent and more inclusive set of preferences, including both rock and symphonic music, is better than the set including only rock music. Others might argue that a rejection of rock music would constitute a further positive transformation. My purpose here is not to defend a particular value system but to show how the notion of transformative value of experiences follows quite naturally upon two beliefs: (1) that some preferences and related value systems are objectively better than others and (2) that the values and preferences held by individuals are altered by experiences they have.

In Chapter 10, I will show how species preservationists often refer to the character-building transformative value of interactions with nature—when they do so, they commit themselves to a belief that certain sets of consumptive preferences are inferior to ones that would be promoted by meaningful interactions with wild species and unspoiled ecosystems. Transformative value attributed to nature does not collapse into or require attributions of intrinsic value to natural objects, provided that the reasons given to justify the objective advantages of the underlying value system are human-related.

Demand value and transformative value represent two types of instrumental value; as such, they must be distinguished from intrinsic value. Critics of anthropocentrism argue that concern for exclusively human interests leaves nature vulnerable to the ravages of ever-growing consumptive demands.[10] But anthropo-

---

[10] As an example of this criticism, see David Ehrenfeld, "The Conservation of Non-Resources," *American Scientist* 64 (1976): 648-656, and *The Arrogance of Humanism* (New York: Oxford University Press, 1981). Despite differences of terminology, Ehrenfeld's argument typifies this line of attack. See Section 11.1 for a discussion of Ehrenfeld's approach.

centrists often note that not all human values are consumptive and emphasize that human aesthetic and moral ideals may be enlisted to limit the exploitation of nature and nonhuman species.[11] They sometimes argue that destruction of other species is adequately proscribed by human ideals and that there is no reason to attribute intrinsic value to nonhuman species.

Anthropocentrism and nonanthropocentrism therefore do not always contradict each other directly. Nonanthropocentrists criticize anthropocentrists for leaving species at the mercy of unlimited human consumptive preferences, while anthropocentrists insist that limits on human consumption can be derived within a richer theory of human value. My distinction between demand and transformative values clarifies this puzzling situation. I will therefore refer to two forms of anthropocentrism. "Strong anthropocentrism" is the thesis that nonhuman species and other natural objects have value only insofar as they satisfy human demand values. "Weak anthropocentrism" is the thesis that nonhuman species and natural systems often have, in addition to demand value for humans, transformative value as well. Weak anthropocentrism is a form of anthropocentrism because it attributes no intrinsic value to nonhuman species.[12] But the range of human values countenanced is much broader than in strong anthropocentrism because other species are considered valuable for their contribution to the formation of human ideals, the sort of ideals that emerge as an individual criticizes and modifies his felt preferences in the struggle to replace them with considered preferences. The concepts and categories here introduced can be represented schematically (see chart). Four approaches to ration-

[11] See, for example, Elliott Sober, "Philosophical Problems for Environmentalism," in *The Preservation of Species*, ed. Bryan G. Norton (Princeton, N.J.: Princeton University Press, 1986). Also see Thomas E. Hill, "Ideals of Human Excellence and Preserving Natural Environments," *Environmental Ethics* 5 (1983): 211-224.

[12] But see Tom Regan, "The Nature and Possibility of an Environmental Ethic," *Environmental Ethics* 3 (1981): 19-34. See especially pp. 25-26, where Regan argues that an ethic appealing to human ideals regarding the treatment of nonhuman individuals and species ultimately requires the assertion that they have intrinsic value. Regan's argument, however, is unconvincing (see Section 11.1), and I believe my trichotomy is far more illuminating than the usual dichotomy.

Intrinsic Values

	A. Of Humans (only human interests are considered)	B. Of Nonhumans (nonhuman as well as human interests are considered)
Demand values	Strong anthropocentrism	Nonanthropocentrism
Transformative values	Weak anthropocentrism	Biocentrism or ontocentrism

ales for preserving species can be observed. Strong anthropocentrists can base their arguments for preserving nonhuman species only on such species' value in fulfilling human demands. Weak anthropocentrists can appeal to human demand values but can also argue that other species reform and enlighten human demands as well as satisfy them. Nonanthropocentrists have available all the principles that strong and weak anthropocentrists do, although they sometimes ignore transformative values. In addition, they can claim that all human uses of nature are circumscribed by the principle that objects having intrinsic value should be preserved. Finally, it would be possible to argue that other species or, more likely, members of other species could value transformations of their own needs. One nonhuman species could have value to another nonhuman species by virtue of its potential to alter the former's "value system." This approach is not popular for good reason. To defend it one would have to defend the view that natural objects and species are conscious because value transformations take place only within consciousness. Arthur Schopenhauer's claim that consciousness and will are distributed throughout all life forms or Alfred N. Whitehead's view that nature is an organism might provide a beginning point for the de-

13

fense of such a view.[13] Such arguments are, however, too speculative metaphysically to have any impact on policy formation,
and so this fourth approach will not be discussed here.

On the proposed taxonomy, then, we are left with a trichotomy
of value approaches to justify the preservation of species. That
policy can be justified solely in terms of human demand values, in
terms of human transformative values, or by appeal to the intrinsic value of nonhuman species. Rationales that appeal to either or
both of the first two categories treat nonhuman species as instrumental to human values. Rationales that appeal to the third type
of reasons must attribute intrinsic value to nonhuman species.

1.3. Scientific Rationales?

One effect of the adopted classification system is to exclude, by
stipulation, purely scientific reasons for preserving species. Is this
unfair and illegitimate?

In one sense there obviously are scientific reasons for preserving species: a scientific understanding of species, their individual
characteristics, their interrelationships with other species, these
and other considerations provide an empirical basis for many,
perhaps all, arguments for preserving species. But I have claimed
that all arguments for preservation rest also on crucial value
premises, and I have chosen to classify them accordingly.

Rationales purported to be scientific fall into two broad categories. One group of arguments derives from the science of ecology.[14] This line of reasoning takes ecology to be composed of generalizations concerning the workings of ecosystems. Arguments

---

[13] See Arthur Schopenhauer, *The World as Will and Idea*, trans. R. B. Haldane
and J. Kemp (London: Routledge and Kegan Paul, 1883), vol. 1; Alfred N. Whitehead, *Science and the Modern World* (New York: The MacMillan Company,
1925).

[14] See, for example, Thomas B. Colwell, Jr., "The Balance of Nature: A Ground
for Human Values," *Main Currents in Modern Thought* 26 (1969): 46-52; Aldo
Leopold, "The Land Ethic," in *A Sand County Almanac* (London: Oxford University Press, 1949), pp. 201-218; and Don E. Marietta, "The Interrelationship of
Ecological Science and Environmental Ethics," *Environmental Ethics* 1 (1979):
195-207.

for preserving species are then thought to emerge from such generalizations—for example, from predictions about the consequences for ecosystems of the extirpation of species. These arguments often suggest that a deep understanding of the science of ecology imbues the individual with a respect for complex, interrelated ecosystems and a knowledge of their beneficial effects. This respect and knowledge manifest themselves in a strong motivation to preserve such systems and their functioning parts.[15] A second group of reasons emphasizes the importance of scientific knowledge, arguing that the many unknown and unnamed species represent a treasure-trove of important information.[16] If such species are lost before they are even catalogued, named, and described, the human race will have lost forever these possibilities of knowledge. In this and the next section I will discuss whether either of these groups of arguments gives a distinctively scientific reason for preserving species.

The conclusion that we ought to save other species is unabashedly hortatory. Normally it is not thought that scientific data and generalizations can, in and of themselves, imply recommendations for action. It is a cornerstone of the modern view of science and ethics that evaluative statements cannot be derived from purely scientific statements. Those who espouse ecological reasons for preserving species give either a general argument against this basic principle or a specific argument that ecology does not conform to the standard, modern paradigm of science.

The general argument proceeds as follows: the modern paradigm assumes that there are purely factual (entirely value-free) statements and that such statements represent the body of scientific knowledge. From these assumptions the standard view concludes that no purely scientific statements can entail an evaluation or recommendation because no statement about the way

---

[15] This approach is exemplified in Paul W. Taylor, "The Ethics of Respect for Nature," *Environmental Ethics* 3 (1981): 197-218.

[16] See Donald H. Regan, "Duties of Preservation," in *The Preservation of Species*; and Alastair S. Gunn, "Preserving Rare Species," in *Earthbound: New Introductory Essays in Environmental Ethics*, ed. Tom Regan (New York: Random House, 1984), pp. 319-320.

things are can, in and of itself, entail a statement about what ought to be the case. Opponents of the paradigm question the first of these assumptions, arguing that in every purportedly scientific statement is hidden a value claim. In the words of John Dewey the split between fact and value is based upon an "abstraction" from true human consciousness.[17] Science has always served human ends and, hence, the statements of science are infused with the evaluative goals of scientists. Dewey says: "When the consciousness of science is fully integrated with the consciousness of human value, the greatest dualism which now weighs humanity down, the split between the material, the mechanical, the scientific and the moral and ideal will be destroyed."[18]

This general line of reasoning is applied to an "ecological ethic" by Don E. Marietta, Jr.:

> To understand the point I am making, we must reject the notion that facts are perceived in bare objectivity (i.e., given by the world, impressed upon a passive mind) while values are only products of our subjective judgment (i.e., produced by the mind). Such a notion of brute, theory-free facts is an obsolete concept, no longer useful in science or the philosophy of science. Both factual and valuational observations of the world are constituted together by consciousness. We seldom perceive value-free states of affairs, and we never perceive facts unmediated by human consciousness. When we make purely descriptive *is* statements, these are abstracted from the world as experienced, the world as we live and participate in it. . . . An ethic founded upon ecology, since it is not derived abstractly from entailment relations between statements, does not deduce *ought* from *is*. It is rather a matter of recognizing the values embedded in our observations of the world, observations in which factual cognition and value cognition are fused, only to be separated by reflection.[19]

[17] John Dewey, *Reconstruction in Philosophy* (Boston: Beacon Press, 1948), p. 174.
[18] Ibid., p. 173.
[19] Marietta, "The Interrelationship of Ecological Science and Environmental Ethics," p. 200.

The specific argument cites special characteristics of ecology that do not fit the normal paradigm. A typical form of the specific argument asserts that any person who studies ecology and comes to understand the complexity, interrelatedness, and fragility of biotic systems will be imbued with a new and different sense of humanity's place in nature. An individual who has this sense of place will recognize the importance of protecting the integrity of ecosystems and will, in turn, understand the importance of saving the species of which they are composed.

Hence, we find William Murdoch and Joseph Connell describing their discipline in the following terms:

> We do not believe that the ecologist has anything really new to say. His task, rather, is to inculcate in the government and the people basic ecological attitudes. The population must come, and very soon, to appreciate certain basic notions. For example: a finite world cannot support or withstand a continually expanding population and technology; there are limits to the capacity of environmental sinks; ecosystems are sets of interacting entities and there is no "treatment" which does not have "side effects" (e.g., the Aswan Dam); we cannot continually simplify systems and expect them to remain stable, and once they do become unstable there is a tendency for instability to increase with time. Each child should grow up knowing and understanding his place in the environment and the possible consequences of his interaction with it.
>
> In short, the ecologist must convince the population that the only solution to the problem of growth is not to grow.[20]

Implicit in this passage is the suggestion that the science of ecology exhibits special characteristics implying recommendations, such as the recommendation that biological diversity should be protected.[21]

[20] William Murdoch and Joseph Connell, "All About Ecology," in *Western Man and Environmental Ethics*, ed. Ian Barbour (Reading, Mass.: Addison-Wesley Publishing Co., 1973), p. 169. Also, see Roger Revelle and Hans H. Lansberg, eds., *America's Changing Environment* (Boston: Beacon Press, 1970), p. xxii.

[21] This view has been developed in a more philosophical manner by Holmes Rolston, III, although he stops short of fully endorsing the position. See "Is There

   Neither the general nor the specific form of the naturalistic objection to the modern scientific world view shows that there are
arguments for preservation not based upon evaluative premises.
They do not, therefore, provide a reason to reject my classificatory system. It is true that, if valid, they show that recommendations can be derived from "purely scientific" premises. But this
does not imply that there are no evaluative premises supporting
these recommendations because these objections insist that scientific data and theory *contain* or *give rise to* evaluative premises.
While it may be more difficult to identify and isolate such premises if the naturalistic objections are true and value assumptions
are hidden in ostensibly factual material, it remains the case that
all recommendations rest upon evaluative premises, however ob-
scured. A difficult analysis of scientifically based recommendations may be necessary in order to ferret out their implicit value
commitments so that the rationale in question can be placed in
my proposed system of categories. This represents an inconveni-
ence but does not undermine the completeness of my classificatory system.

   A second type of argument that has been described as scientific
treats each species as an important object and source of knowledge. The extinction of a species is said to be analogous to the
burning of all copies of a book.[22] The genetic, physiological, and
behavioral information presented in a species will be forever lost
as an object of scientific study if that species is extinguished, given
that studying preserved specimens or fossils is a poor substitute
for studying living and behaving creatures. The extinction of a
species represents a loss in the development of knowledge because once those objects of knowledge disappear, we will never
retrieve the opportunity to understand fully a unique living thing.

---

an Ecological Ethic?" *Ethics: An International Journal of Social, Political and Legal Philosophy* 85 (1975): 93-109. In later essays Rolston has moved closer to a
full acceptance of such a line of reasoning: "Can and Ought We to Follow Nature?" *Environmental Ethics* 1 (1979): 7-30, and "Values Gone Wild," *Inquiry* 26
(1983): 181-207.
   [22] See, for example, Norman Myers' quotation of Thomas Lovejoy in *The Sinking Ark* (Oxford: Pergamon Press, 1979), p. 7. Also see Holmes Rolston, III, "Duties to Endangered Species," *Bioscience* 35 (1985): 718-726.

Does this line of reasoning represent a purely scientific argument for species preservation? Obviously not. References to "the importance of scientific knowledge" all depend on some nonscientific source of value.

First of all, science serves human demand values. When science discovers that a particular species contains an unusual chemical defense against predators (for example, that the venom of the Malaysian pit viper acts by thinning the blood of its victim),[23] this piece of knowledge may simply be an interesting fact. Or it may turn out that if the chemical is extracted or simulated it will act as a life-saving anticoagulant when administered with care to humans.

In the latter case the scientific knowledge is clearly instrumental to the protection of human life. The preservation of the viper will have preserved an object of scientific knowledge, and the knowledge of that object will, in turn, have served human interests. But these human interests do not transcend the realm of human demand values. The value of science, in this case, is to make possible an advance in medical technology that enhances and protects human life. The preservation of the species in question serves no distinctively scientific value: its value is instrumental to the fulfillment of human demand values.

But what about cases where no technological advances result from the study of a particular species? Suppose the discoveries in question prove merely to be fascinating facts about nature. Is the value of such discoveries purely scientific—providing the basis of an argument that the preservation of such a species serves a purely scientific value? First, it should be noted that such discoveries might be indirectly instrumental to inspiring researchers who study similar forms that do turn out to be useful for technological reasons. Or the data obtained might plug a gap in scientific knowledge and allow an important theoretical breakthrough. But in each of these cases, if the benefit derived ultimately lies in the species providing a steppingstone to further

[23] Norman Myers, *A Wealth of Wild Species: Storehouse for Human Welfare* (Boulder, Colo.: Westview Press, 1983), p. 119.

technologically useful knowledge, the reason in question can again be assigned to the realm of human demand values. Leaving aside these cases of derivative value, what could be said if a piece of scientific information proved merely interesting, with no foreseeable direct or indirect payoff in technological advances?

Many researchers, describing themselves as pure scientists, argue that they study other species and natural systems for their inherent interest alone. Without disputing these motives, it still does not follow that the preservation of species for research entails that species have purely scientific value. Knowledge need not be pursued for the sake of technological advances. When it is not, however, it is pursued for its own sake—for advances in the understanding of the beauty and interest of nature—and this, too, is a valued human goal.

The stipulation, therefore, that arguments for preserving species be organized according to central value premises does not unduly narrow the field of possible arguments. Arguments sometimes thought of as scientific in nature fit readily into the proposed categories, when the necessary value premises implicit in them are made explicit.

The Endangered Species Act functions as a "flagship" of the environmental movement. It also functions, from time to time, as a battleground on which economic interests vie with preservationist interests. To the extent that the act is justified by economic reasons, the battles it precipitates may well be no more than competitive struggles for larger pieces of the economic pie. If the act serves only the narrowly commercial interests of the nation, then it becomes, in a sense, the tool of special interests within the economic community.

But many observers and participants view the battles involved as more than questions of distributive fairness among competing economic interests. The feeling persists that the act, at least at times, sets productive interests in resources, in general, against some other interests. If this is true, and if the act is on the whole justifiable, then there must be some justification for the act that does not derive from its protection of resources for fulfilling hu-

20

man desires. That insight has here been embodied by classifying reasons for preserving species according to a basic trichotomy of human demand values, human transformative values, and intrinsic values of nonhuman species.

Because the proposed classificatory system focuses attention on this trichotomy, it raises many interesting value questions. This focusing is made possible because the system classifies reasons or arguments according to their underlying value premises. I have argued that it can comprehend all arguments, including the ones sometimes called scientific. I believe it also encourages a useful discussion of how demand values relate to nondemand ones. It thereby encourages discussion of comparative rationales in a way that at least offers the hope that light will be shed on questions of exceptions, implementation, and priorities.

Consequently, I will devote Parts A, B, and C of this book to examinations of the various rationales for preserving species, with the trichotomy of demand values, intrinsic values, and transformative values serving to delimit the generic categories of rationales. Part A examines the strength of arguments based only on human demand values. Because demand values provide, most basically, the motives explored in economic analyses, it will be important to consider whether, and in what manner, these values can be accurately expressed in such analyses. The reasons for and against attributing intrinsic value to nonhuman species are the subject of Part B. The argument of these chapters will be expressed in mainly philosophical terminology. Since the arguments of the second part will prove inconclusive, Part C will explore the possibility of defending species on the basis of their transformative value to humans. Part A, taken alone, represents the strongly anthropocentric approach. If the arguments in Part A are supplemented with affirmations of intrinsic value as discussed in Part B, the result is a nonanthropocentric position that still recognizes the demand value of other species to humans. If Part A is supplemented by considerations from Part C alone, then the approach is weakly anthropocentric. It would, of course, be possible to accept arguments from all three parts, thereby resting the case for species preservation on a very broad base of considerations. Part D, not concerned directly with rationales, draws upon the conclusions of the earlier parts and discusses the problem of priorities— What ought to be done when there are insufficient resources available to save all species?

A

DEMAND VALUES AND SPECIES PRESERVATION

T W O

DEMAND VALUES
AND ECONOMIC ANALYSIS

2.1 Benefits and Dollars

What is the value of species? In evaluating public works projects that affect them, it is sometimes proposed that decisions can best be made by calculating in dollars the value of species whose survival chances would be affected. Once assigned, these dollar values can be factored into economic analyses informing decisions about the project. If, for example, the project threatens the future existence of a species, the dollar value of that species should be considered one of the project's costs. If the project protects a species, that value should be added to the other benefits involved. It is often assumed that a dollar value, so assigned, would represent the benefits the species offers humans. This assumption must now be evaluated.

It is crucial to note, first of all, that any method of computing dollar values represents a stipulated replacement for the very general, imprecise everyday concept of a "benefit." Much as physicists use the precisely measurable concept of mass to replace the common concept of weight, economists provide a method for computing dollar values to replace the ordinary sense of benefits. Recognition of this fact implies two further points: (a) the various possible methods of computation proposed may lead to extremely different outcomes and, consequently, these methods may approximate the ordinary concept of benefits with differing degrees of accuracy; and (b) there is a justifiable fear that any proposed method of computing dollar values will systematically bias decisions by ignoring or miscalculating certain qualitative values that species have, thereby narrowing the concept of benefits and systematically undervaluing species.

This fear derives from three main sources. First, the framing of the value question itself affects the outcome. The normal method of assigning dollar values to individual species, examined one-by-one, distorts value assignments. The extent of this distortion will be the subject of Section 2.2.

Second, many values of species cannot be assessed with any precision. The scientific knowledge necessary to quantify them is simply not available. At best, we can compute the known commercial value of identified species. But once we look to future unknown uses we are on shaky ground. Matters grow still worse when we turn to values derived indirectly through ecosystem services. This second fear is taken up in Section 2.3 and in Chapter 3 and Chapter 4.

Finally, by attempting to quantify all benefits, economic analyses must treat all values as demand values. To assign a dollar figure to any object is to treat it as a market commodity. But it is a truism that not all objects and processes are market commodities. Some of them have no price; others are priceless. They are "external" to markets. Economists have, accordingly, proposed that they be assigned a price through hypothetical market techniques, such as the willingness of consumers to pay to preserve such objects or processes. In Section 2.4, I will argue that these techniques necessarily distort the value of species because it is impossible to determine how to evaluate accurately some values of species.

Before developing these arguments, it is necessary to mention briefly the various types of demand values that are attributed to species and factored into the standard calculations of economic evaluations. Since detailed discussions of these values abound, I will only mention the most important categories.[1]

---

[1] For more detailed discussions, see, for example, Anthony C. Fisher, "Economic Analysis and the Extinction of Species," Report No. ERG-WP-81-4 (Berkeley, Calif.: Energy and Resources Group, University of California, November 1981); Alan Randall, "Human Preferences, Economics, and the Preservation of Species," in *The Preservation of Species*, ed. Bryan G. Norton (Princeton, N.J.: Princeton University Press, 1986); Paul Ehrlich and Anne Ehrlich, *Extinction:*

Randall lists four general categories of economic values associated with species preservation: (1) use values; (2) option values; (3) quasi-option values; and (4) existence values.[2] Use values include amenity uses as well as extractive ones. Many authors have emphasized the many and surprising ways in which human beings draw upon biotic resources to protect and improve their lives. These include gathering wild species for food, domesticating wild species of plants and animals for agricultural products, using naturally occurring products in industry, deriving pharmaceutical products for the improvement of health, and burning of organic products for fuel. But use values also include amenities such as aesthetic enjoyment of parks and zoos and the naturalist's interest in other species as a source of study and enjoyment. Many writers have noted the psychological advantages of living in those cities in which built environments are interspersed with parks and of retreats to nature as a source of solitude and literary inspiration.

Option values are defined derivatively upon use values. Other species have value even if they are not currently used because they could be used in some of the ways just listed. If a species is lost, the possibility of deriving some use from it is lost with it. This possibility must be represented as a value that species have to humans. Quasi-option values are option values enhanced by the expectation that growth in knowledge will find as yet unknown uses for species.[3]

Existence values are defined independently of use and, therefore, it is questionable whether they can be construed as demand values at all. I will argue that existence values must be understood as nondemand values and that it is a conceptual error to try to quantify them. If this case is persuasive, it follows that economic analyses must be flexible enough to include qualitatively expressed values.

The Causes and Consequences of the Disappearance of Species (New York: Random House, 1981); and Norman Myers, *A Wealth of Wild Species: Storehouse for Human Welfare* (Boulder, Colo.: Westview Press, 1983).

[2] Randall, "Human Preferences."
[3] Ibid.

2.2 Framing the Question: Alternative Approaches to Economic Analyses

Ecologists and environmentalists often insist that there are important "noneconomic" values or that economic analyses are inadequate means to evaluate decisions affecting the environment.[4] In some circles "benefit-cost analysis" is a term of derision. The meaning of these beliefs and attitudes is not always clear because there is no single way to formulate expressions of economic value. One purpose of this book is to give clear expression to, and to evaluate, these antieconomic sentiments as they are voiced regarding the evaluation of endangered species.

Because crucial concepts are used ambiguously, it will help to enforce, even against common usage, some terminological conventions. I will distinguish between "benefit-cost analyses" (BCAs) and "economic analyses." When I refer to BCA's, I will refer specifically to analyses that countenance values only if they are stated in monetary terms. BCA's are here characterized by three assumptions. (1) Quantitative values can be rationally assigned to costs and benefits resulting from proposed courses of action. (2) These quantitative values represent individual preferences, measured by the individual's behavior in markets. If there is no market for a good or amenity, value will be assigned hypothetically, by determining what the individual would be willing to pay (WTP) for an increment of it or what compensation the individual would be willing to accept (WTA) as compensation for a decrement of it.[5] (3) Public decisions would be determined by the outcome of BCAs. That course of action should always be adopted which produces the most favorable ratio of benefits to costs, calculated without regard to whom they accrue.[6]

---

[4] See, for example, Aldo Leopold, *A Sand County Almanac* (London: Oxford University Press, 1949), p. 224; David Ehrenfeld, *The Arrogance of Humanism* (New York: Oxford University Press, 1981), Chapter 5; and Ehrlich and Ehrlich, *Extinction*, Chapter 3.

[5] Randall, "Human Preferences," Section II.

[6] Assumption (3) implies that a BCA, once completed, should determine public policy decisions. By making this assumption central to "the BCA approach," I no

This criterion of costs and benefits would determine public decisions on the BCA approach as I have characterized it. By contrast, I will use the phrase "economic analyses" more broadly to include approaches that favor quantified preferences when possible but that recognize qualitative values as well. Economic analysts (as I use the phrase) believe that quantified preferences are sometimes overridden by unquantifiable values, and they expect to balance quantified data on preferences against other social priorities.

BCAs became popular among government administrators in the 1940s and 1950s because postdepression policy makers believed that government activities should encourage, rather than stifle, economic activity.[7] When resource economists began taking species preservation, aesthetic aspects of the environment, and other less clearly economic values seriously in the 1960s, they faced an entrenched bias in favor of more narrowly defined commercial values that are readily quantified in BCAs. Responding to this existing situation, some resource economists employed the WTP and WTA criteria of value with some success. They accepted the challenge of assigning hypothetical values to the whole range of social values and showed that preservation options often fare surprisingly well against development options when a wide range of values is computed. Faced with the alternatives of participating in an established decision process or becoming conscientious objectors and forfeiting any voice, they chose to participate.

---

doubt use that label more narrowly than some writers. Randall, for example, describes the "BCA criterion" in a manner consistent with assumptions (1) and (2) but remains skeptical about using it as a public decision rule. He argues that it must be superimposed on the body of existing law so that maximizing benefits over costs should be used only to choose among alternatives consistent with the basic law of the land. He also suggests that other considerations, such as those of equity, should override the criterion in some circumstances. I feel justified in narrowing usage for the sake of clarity because many benefit-cost analysts would advocate its use as a public decision rule. As for existing legislation and programs to redistribute wealth, they would insist that these should likewise be judged by the same criterion. Accordingly, advocates of the BCA approach as I understand it are committed to assumption (3) as well as to assumptions (1) and (2).

[7] Mark Sagoff, "Ethics and Economics in Environmental Law," in *Earthbound: New Introductory Essays in Environmental Ethics*, ed. Tom Regan (New York: Random House, 1984), p. 148.

Nevertheless, it remains an important question whether the three assumptions of the BCA approach can be justified. If they cannot, environmentalists and resource economists should join to protest arbitrary administrative decisions that offer no alternative but to assign quantified values. Resource economists, speaking as economic analysts, could register this protest by insisting on the importance of qualitative values in public decision making. They must state clearly that not all questions can, given the present development of their art, be given realistic and reasonable quantified representation. Thus, even while they participate in the BCA process and accept its methodological desideratum that values be quantified whenever possible, these resource economists can also note the limitations of its methodological assumptions and caution against the distorting effects of following them slavishly.

Can the BCA approach provide an accurate assessment of the value of endangered species while adhering to its three characteristic assumptions? When applied to questions of species preservation, the BCA approach assesses the value of an individual species and, if several species are involved, sums the relevant individual assessments.[8] With this approach it is assumed that any value the species has relates to particular, specifiable ways in which it is useful or in which it is enjoyed. No initial presumption in favor of species preservation is built into the analysis. This piecemeal approach, however, is not the only possible option. If a more diverse environment is, other things being equal, preferable to a less diverse one, then a positive value should be initially assigned to every species, regardless of the ability of a BCA to list specific values for it. While this value might be overridden by negative instrumental values (as has been argued for the smallpox virus) or by greater values to be derived from the project that threatens the species, this approach shifts the burden of proof by requiring that the assessment process begin with the presumption that, other things being equal, a species has positive value.

I suggest, opposing BCAs, that a quite different evaluation

---

[8] See Fisher, "Economic Analysis and the Extinction of Species."

might be forthcoming if the question is framed, "What is the value of a species considered as a unit of biological diversity?" and not "What is the value of this particular species?" If biological diversity itself contributes to the fulfillment of human preferences, it thereby has demand value, generating a prima facie reason for protecting all species. It might be thought that I overemphasize this difference of approach, because surely it is possible to add a category to a standard BCA that would represent this prima facie value. This value would then be added in every case to the values specified for the individual species under examination.

Two responses are in order. First, whether or not this is possible, it is not normally done. The species-by-species evaluations currently prepared do not take the general diversity factor into account. Second, I will argue below that it is impossible to assign any realistic dollar amount to such a prima facie value. If that argument is sound, then the proposal to incorporate the prima facie value into BCAs would overturn the first assumption of that approach—that all items of value can be assigned a rationally supportable, quantified monetary value.

Determining this general value of a species amounts to determining the value of a unit of biological diversity. Or, to put the point in a different way, it is to assess the value of a species, without regard to the individual or populational characteristics of the species. The concern, then, is with diversity itself. Biologists and ecologists distinguish three concepts of diversity, however, and it will be useful to survey them before continuing the argument of the chapter.[9] Insofar as an ecosystem can be viewed as a separate and (partially) closed system occupying a delimited space, it is possible to measure its *within-habitat* diversity.[10] This type of diversity could be designated in a rough-and-ready way by simple species counts, although many ecologists prefer measures that

---

[9] See R. H. Whittaker, "Vegetation of the Siskiyou Mountains, Oregon and California," *Ecological Monographs* 30, no. 3 (July 1960): 320; Robert H. MacArthur, "Patterns of Species Diversity," *Biological Review* 40 (1965): 510-533.

[10] MacArthur, "Patterns of Species Diversity," pp. 515-516.

also indicate relative abundance of species.[11] A system composed of one dominant species and n rare species is considered less diverse than a system composed of the $n + 1$ species with more even distributions. Within-habitat diversity also depends on the degree of difference among the species existing in the system. Two species from the same family add less to diversity than do two unrelated species.

Diversity also has a *cross-habitat* dimension.[12] An area is considered more diverse if it contains a number of very different systems; cross-habitat diversity indicates the heterogeneity existing among the various habitats and ecosystems in an area. When foothills of a mountain range approach the sea, habitats are compressed into narrow ribbons along the shore, as elevations vary rapidly. Such areas have a high degree of cross-habitat diversity.

Total diversity, while not independent of the first two types, differs because it is a measure of species diversity for a geographical area. Within-habitat diversity measures the variety of species for a delimited system. Cross-habitat diversity refers to the variety of habitats in a geographical area. Total diversity combines these two measures, referring to the variety of species in all of the habitats existing in an area. The area can be defined as an environmentally bounded space (an island, say, or an area lying between a river and a mountain range) or, alternatively, the boundary could be political (a state or county line), in which case it is irrelevant ecologically but perhaps relevant to policy questions. Again, species counts may serve as a rough indicator of total diversity,[13] but it must be remembered that other factors such as evenness of distribution and the phylogenetic distance between species may affect total diversity in the same way that they affect within-habitat diversity. The total diversity of an area is clearly affected by the within-habitat diversity of the systems existing there and by the diversity of structure in those habitats.

Total diversity is most significant as an indicator of the variety

[11] Ibid., pp. 511-515.
[12] Ibid., pp. 522-523.
[13] R. H. Whittaker, "Evolution and Measurement of Species Diversity," *Taxon* 21 (1972): 232.

of species available to colonize newly disturbed areas. Even very rare species may embody adaptations that will make them successful competitors for emerging niche space, so contribution to total diversity is not necessarily limited by rarity. It may, of course, be true that a species is rare because it is specialized to a very narrow habitat. But its specialized adaptations may also make it an especially intense competitor for a new space, provided that space fulfills the niche requirements of the species.

Although within-habitat diversity and cross-habitat diversity are more commonly discussed by ecologists, total diversity is the relevant concept in discussing endangered species policy. The local extirpation of a species from a particular ecosystem need not be cause for alarm. Natural successional changes in that system may involve an ecologically predictable shift in species composition. As long as the species in question survives in other nearby systems, it will remain available to recolonize systems from which it disappears if, for whatever reason, the locally altered system is returned to an earlier stage of succession and the species becomes a viable and appropriate part of it. If species preservationists show concern for each threat of extirpation from a particular habitat, their efforts would be applied too broadly: as long as the species continues to exist in other ecosystems within the locale, it need not become an object of concern. But preservationists should not show concern for global extinctions only. If a trend develops whereby a species disappears from more and more particular habitats and is in danger of being lost from the larger geographic area, this should be of concern, as the geographic contraction of the range of a species makes it more vulnerable to global extinction. Geographically separate populations contribute to the diversity of the gene pool and protect against catastrophic events like epidemics of disease or outbreaks of predators that could wipe out entire local populations.

Species preservationists, then, should direct their concern at areawide trends—attempting to maintain representation by species in as many geographic areas as possible. The concept of total diversity corresponds to this appropriate level of concern. By protecting the total diversity of an appropriately sized geographical area, species preservationists protect species from approaching dangerous stages in which one or a few isolated populations represent the only hope, without showing premature alarm at the natural disappearance of a species from a particular system undergoing changes in successional development. The general value of a species—the value it has independent of its individual and populational characteristics—is represented by its contribution to the total diversity of the area in which it resides. The question of Part A can thus be framed: What are the advantages (expressible as demand values) to humans of living in a geographic area characterized by great total diversity?

Although the relevance of total diversity for species preservation has seldom been noted explicitly, a number of naturalists and ecologists have implicitly fallen into the intuitive use of this concept when they discuss the importance of species preservation. For example, this seems to be the motivation behind Aldo Leopold's remark that "to keep every cog and wheel is the first precaution of intelligent tinkering."[14] And when Charles Elton discussed "The Conservation of Variety," he argued for the importance of a patchwork environment composed of mixed fields, hedgerows, and roadside verges as refuges for a maximum number of species. This concern for diverse habitats in an area could only indicate a concern for total diversity.[15]

For these reasons I suggest that the goal of preserving species be conceived as the goal of protecting total diversity, first of areas, and, by extension, of whole continents and the world itself. Each species is thus conceived as a unit of total diversity, and responsible decision processes responding to threats to the existence of

[14] Aldo Leopold, "Conservation," in *Round River: From the Journals of Aldo Leopold*, ed. Luna B. Leopold (New York: Oxford University Press, 1953), p. 147.

[15] Charles S. Elton, *The Ecology of Invasions by Animals and Plants* (London: Methuen, 1958), pp. 154-159; also see Gerald Lieberman, "The Preservation of Ecological Diversity—A Necessity or a Luxury?" *Naturalist* 26 (1975): 24-31; J. W. Humke et al., *Final Report: The Preservation of Natural Diversity: A Survey and Recommendations*, report prepared for the U. S. Department of the Interior by the Nature Conservancy, 1975; President's Council on Environmental Quality, *Global 2000 Report to the President: Entering the Year 2000*, U.S. Government Document, 1980.

a species must take into account the value of each unit of total diversity. This is not, of course, to deny that particular species have special values deriving from their populational or individual characteristics. The point is that if total diversity is valuable, then every species should be accorded some positive value. Those species that have special value should have that value augmented by the same factor. In this manner it may be possible to assign comparative values to species even though, as I will argue later, it is impossible to assign meaningful and nonarbitrary dollar values to them.

Two alternative approaches to decisions concerning endangered species have emerged in the literature of economics. These have been dubbed the "RFF approach" and the "SMS approach."[16] The former approach derives largely from the work of economists connected with Resources for the Future, especially that of Anthony Fisher, John Krutilla, and V. K. Smith and applies a modified form of benefit-cost analysis.[17] The latter approach, designated as the "safe minimum standard," originated in the work of S. V. Ciriacy-Wantrup and has been further developed by Richard Bishop and Oliver Ray Stanton.[18] It concentrates on the social costs of species extinctions, adopting the rule: "Avoid extinction unless the social costs of doing so are unacceptably large."[19] Its approach to species extinction parallels Cir-

[16] Randall, "Human Preferences."

[17] Fisher, "Economic Analysis and the Extinction of Species"; also see V. K. Smith and J. V. Krutilla, "Endangered Species, Irreversibility, and Uncertainty: A Comment," *American Journal of Agricultural Economics* 61 (1979): 371-375; J. R. Miller and F. C. Menz, "Some Economic Considerations in Wildlife Preservation," *Southern Economic Journal* 45 (1979): 718-729; and Charles Ploudre, "Conservation of Extinguishable Species," *Natural Resources Journal* 15 (1975): 791-798.

[18] S. V. Ciriacy-Wantrup, *Resource Conservation: Economics and Politics* (Berkeley and Los Angeles: University of California Division of Agricultural Sciences, 1968); Richard Bishop, "Endangered Species and Uncertainty: The Economics of a Safe Minimum Standard," *American Journal of Agricultural Economics* 60 (1978): 10-18; Richard Bishop, "Endangered Species: An Economic Perspective," *Transactions of the 45th American Wildlife and Natural Resources Conference* (1980), pp. 208-218; Oliver Ray Stanton, untitled remarks delivered at the American Academy for the Advancement of Science Meetings, Washington, D.C. (January 4, 1982); also see Randall, "Human Preferences."

[19] Bishop, "Endangered Species and Uncertainty," p. 10.

iacy-Wantrup's general approach to resource preservation: protecting resources is always preferable, other things being equal, because the protectionist approach preserves future options.

While these approaches converge in important ways, I will defend the SMS approach because it embodies the central assumption that all species have value. The arguments of this and the next two chapters provide a justification for that assumption. My arguments therefore complement the SMS approach, justifying its central assumption and showing that, given that assumption, a cautious approach designed to avoid extinctions is the most rational approach to decision making regarding species.

Since both approaches are described by Randall and others, I need only summarize their main features, highlighting points relevant to the arguments of this and the next chapter.[20] The RFF approach begins with the assumption that decisions affecting the survival of a species are to be treated as questions of economic efficiency, in which benefits and costs of various courses of action should be computed and compared. Such comparisons are not normally thought to be definitive, partly because they may contain vague value assignments and partly because factors other than efficiency may affect public decision making. But special circumstances are associated with decisions affecting endangered species that make these comparisons of benefits and costs even less definitive. In Randall's terminology, option values, quasi-option values, and existence values will often be more important than use values. And these types of values are progressively less susceptible of definitive quantification.[21] Fisher, for example, notes that conventional computations of costs and benefits will be biased in favor or development over species preservation unless they are modified to take into account (a) the irreversibility of species extinction; (b) the uncertainty of whether and how a development project will affect a species; and (c) the uncertainty of whether and in what ways the lost species might provide resource benefits.[22]

[20] Randall, "Human Preferences."
[21] Ibid.
[22] Fisher, "Economic Analysis and the Extinction of Species."

That is, besides all the normal difficulties of basing decisions on BCAs, decisions about endangered species have a problematic asymmetry: the decision not to go ahead with a development project is reversible; the decision to let a species go extinct is not. But only about 1 percent of species have received any screening for potential value.[23] If the species does have utilitarian value based upon its populational characteristics and characteristics of its members, then this will most likely not be known. This point is well expressed by Bishop when he quotes an unidentified person as saying that "resources are not, they become."[24] The usefulness of a species may wait upon changes in human tastes and preferences, changes in income levels, developments of knowledge and technologies for using the species, and changes in public policies.[25] But if the species is extinguished before it is examined for usefulness or before such changes can take place, the resulting losses will never be known.

Fisher suggests, then, the following modifications of conventional CBAs: (1) choosing a relatively low rate of discount (the rate at which benefits and costs are diminished across time), because values of many species will emerge in the future; (2) calculating the losses in extractive values (for species of known value) caused by diminution of their populations and operations; (3) calculating the probabilities that the species will actually become extinct and factoring such losses into the analysis diminished by the possibility that such extinction may not occur; and (4) calculating the probability that a species of no known value will prove useful, calculating the value it might have, and diminishing that value by the probability that it will not become extinct as a result of the proposed development project. These steps will, according to Fisher, redress the bias in favor of developmental options and make BCAs more likely to favor preservation options over developmental options.[26]

But Fisher is well aware of the pitfalls involved in carrying out

---

[23] Ibid., p. 10.
[24] Bishop, "Endangered Species and Uncertainty," p. 11.
[25] Ibid.
[26] Fisher, "Economic Analysis and the Extinction of Species."

his recommendations: "What about prospects for estimating benefits for as yet untested and perhaps even unknown species, on the basis of information about some currently useful ones? This seems a hopeless task, yet it is just what we must do if we wish to oppose quantitative estimates of benefits to those which will surely be forthcoming for costs of preserving species and their habitat."[27] Because of these pitfalls, advocates of the RFF approach recommend giving the benefit of the doubt to preservation in cases where computed costs and benefits are apparently balanced or nearly so. But the operational value of such a suggestion is questionable. If available figures representing benefits of preservation are admittedly speculative and grossly inaccurate, how is it possible even to judge when costs and benefits are in fact nearly balanced?

The major difficulties arising on the RFF, modified CBA approach appear on the benefits side of the ledger. Accordingly, advocates of the SMS approach eschew attempts to quantify the benefits of species, choosing instead to focus on the costs of preservation options. Relying on their central assumption that every species has great but unquantifiable value, they argue that species should be saved as long as the costs are tolerably low. This approach represents, in essence, a modification of the minimax rule: choose the alternative that minimizes maximum possible losses.[28] The unmodified rule always avoids the worst outcome, whatever the costs of doing so, but the modified rule admits (in the phrase "as long as the costs are tolerably low") the possibility that in the face of high costs society might choose a small risk of serious negative consequences.[29]

The SMS approach has two difficulties. First, its criterion is only as plausible as its central assumption that every species has significant positive value.[30] Second, it is difficult to give concrete meaning to "tolerably low" costs. In the next chapter I will provide an argument linking economic and ecological theory in a

[27] Ibid., p. 13.
[28] Bishop, "Endangered Species and Uncertainty," p. 13.
[29] Ibid.
[30] Randall, "Human Preferences."

way that resolves the first of these difficulties. In the process the second will be somewhat ameliorated, as a convincing argument that there is significant positive value in saving any species supports a relatively high threshold for tolerating costs. Thus, while it is possible to establish that costs can be quite high before they become intolerable, this crucial phrase will remain vague because it depends, at least in part, on the extent to which society is willing to sacrifice present income for future options.

Before turning, in the next chapter, to the development of the ecological/economic argument for the central assumption of the SMS approach, I will detail, in the next two sections, some of the most serious difficulties that undermine attempts to assign dollar values to individual species, thereby providing reasons not to accept the three stated assumptions of the BCA approach and reasons not to be sanguine about the usefulness of BCAs in evaluating decisions affecting endangered species.

2.3 *Failures in Quantification*

At first glance economic evaluation of a species that has some known and well-established commercial value provides the best illustration of how to assign dollar amounts to the value of a species. For example, it should be possible to compute the total dollar amount that consumers pay for the products created from a species of whales. A dollar figure computed in this manner has an appealing authority. But at second glance the authority proves entirely unfounded and ephemeral, as the computed figure represents the value of the species only on the assumption that its current uses are the only and the best uses possible.

Once it is recognized that any species, however useful for a given purpose, remains a candidate for new and different uses, it becomes clear that the recognized commercial value of a species is merely the *known* value of that species. Insofar as the species remains a candidate for new and undiscovered uses, the dollar figure that emerges from these computations is subject to all the uncertainties attending the assignment of dollar values to a species not yet examined for commercial usefulness. The (uncomputable)

39

likelihood that other uses for the species in question would have been discovered during the development of its current use is countered by another (equally uncomputable) likelihood that species which prove useful for one purpose are thereby improved candidates for new uses. Thus, to the extent that dollar amounts assigned option values and quasi-option values are less accurate than actual commercial values, these very same inaccuracies apply to the dollar amounts assigned species with current commercial uses.

These inaccuracies of computation are staggering. In a presentation before the American Academy for the Advancement of Science Symposium on Estimating the Value of Endangered Species of Plants, Norman Farnsworth attempted to compute the potential pharmacological value of species that are projected to go extinct by the year 2000.[31] By the time all these shaky estimates were compounded, the final figures stated could only be described as speculative.[32] But even these speculative figures are guided by a desire to rely on the "hardest" data available: herbal preparations sold without prescriptions were ignored entirely because the data concerning them were inaccessible.[33] All this uncertainty is generated in a discussion of one possible area of commercial use (medicine) of one type of species (plants). No mention is made of possible multiple uses for the same species and the option and quasi-option values that would arise if a comprehensive value were to be assigned for all possible use of a given species.

These option values and quasi-option values raise special difficulties, especially regarding the distribution of benefits over time.

[31] Norman R. Farnsworth, "The Potential for Plant Extinction in the United States on the Current and Future Availability of Prescription Drugs," paper read at Symposium on Assessing the Economic Value of Plant Species, American Association for the Advancement of Science, Washington, D.C., January 4, 1982.

[32] I refer, in particular, to the manner in which Farnsworth chose the number of plants on the planet. This was only one among a host of estimates involved in determining the number of species affected, including a very questionable assumption that the extinction rate in the United States will be equivalent to that in the rest of the world. All these uncertainties were then compounded with further ones concerning the values of prescriptions, the likelihood that any given species will prove valuable, the amount of scientific examination of species undertaken, etc.

[33] Farnsworth, "The Potential of Plant Extinction."

Many species not now known to be useful will become so in the future. But how far into the future must we look? The irreversibility of the loss of a species implies that the effects will be eternal. Adopting a relatively low rate of discount, as Fisher advocates, lessens but cannot eliminate this inaccuracy. Economic benefits must be computed for finite times. The benefits to future individuals who will have economic needs beyond the time frame chosen are necessarily ignored.

These problems are particularly acute in trying to compute the benefits derived indirectly from species through their contribution to ecosystem services. No ecologist would claim to understand the ecological role of any given species well enough to give an accurate assessment of its contribution to the ecosystems in which it participates. Since humans clearly derive services from ecosystems, some value would be assigned each species that contributes to the system providing the services. But ecologists cannot furnish economists with the information necessary to distribute these benefits among species, even if the value of the ecosystem service could be accurately measured and quantified. Further, species contribute to ecosystems in ecological and evolutionary ways that have no direct or immediate, and certainly no measurable, effect on human benefits. But these contributions are essential to the long-term health of systems and even to the future of the human species. While it is often the case that decisions must be made without full information, the present case is so extreme as to call into question the value of assigning quantified values at all. This problem will be discussed in detail in Chapter 3 and Chapter 4.

2.4 Existence Values

Existence values are defined independently of use.[34] The existence value of a species is the value one attaches to the knowledge that the species exists, even though no use is known or contemplated

[34] Randall, "Human Preferences."

41

for it.[35] Since use values are defined broadly to include aesthetic and amenity uses, existence values are not, in any sense, consumptive in nature. They are not, therefore, market values in the usual sense of that term. Species that have only existence value have, by hypothesis, no use. No one will pay to use them, and no market value can be assigned in the straightforward manner in which objects of use can be assigned market value. It would appear, then, that existence values cannot be included in a BCA.

Undaunted by the seeming finality of this argument, benefit-cost analysts introduce hypothetical techniques, WTP and WTA, for putting a price on existence values. They ask individuals who attribute existence value to a species how much they would be willing to pay to retain the species and thereby keep its existence value intact. Or, alternatively, they ask how many dollars the individual would freely accept as compensation for the loss of the valued species. In this manner species that seem useless, that have no use in the broadest sense of that term, are given demand value. Their demand value equals what preservationists would pay to keep them in existence. They thereby become an object of felt preferences in an extended sense.

Indeed, they are given value in the *only* sense that an object *can* be given value in this system. Given the assumptions of the BCA approach, all values must be demand values because the BCA's first assumption requires that values be assigned according to the single scale of preference satisfaction. Only that scale can be calibrated in terms of dollars. For the sake of the analysis, preferences are accepted as expressed in real or hypothetical markets: in my terminology they are felt preferences. Existence values can be given accurate representation on a BCA only if all values can be accurately reduced to (expressed correctly in terms of) demand values.

To show that some values cannot be so reduced, I will consider two possible existence values. Suppose, first, that existence values attributed to species are really intrinsic values, and, second, that existence values are transformative values. That is, suppose that

[35] Ibid.

upon further questioning it turns out that a respondent values a species that has no use because she attributes intrinsic value to it and feels an obligation to protect it. Or, suppose that the respondent clearly states that she values an otherwise useless species because experience of it may transform someone's felt preferences, making them less materialistic and consumptive.

In the first case, the advocate of the BCA approach must argue that the intrinsic value attributed to the species in question is equal, in dollars, to the amount those who value it would pay to avoid its extinction. But this is a simple confusion. The intrinsic value of any object is by definition the value it has independent of its contribution to human values. To see that the BCA reduction is confused, suppose that terrorists are holding two human hostages for ransom. Suppose, also, that one of the hostages is cantankerous, evil, ugly, has no family or friends, and is loathed by everyone who knows him. The other hostage is attractive, kind, gentle, talented, has a rich family that dearly loves him, and is universally liked. The family of the second hostage comes forward immediately with one million dollars to assure his release, while no one volunteers anything to save the first. By the same reasoning that reduces the intrinsic value of species to what people are willing to pay to save them, the second hostage has one million units of intrinsic value while the first has none. But however ethical theorists interpret the intrinsic value of human individuals, they all agree that this value does not depend upon the individuals' attractiveness to other humans. A similar argument shows that the intrinsic value of species cannot be reduced to demand value by hypothetically inferring felt preferences. The willingness of humans to pay to preserve a species is irrelevant to its intrinsic value.[36] Dollar figures assigned to attributions of intrinsic value cannot reflect that kind of value.

In the second case, where the respondent attributes transformative value to a species, a similar conclusion follows. Again, the claim that transformative values can be expressed as demand val-

[36] See Tom Regan, "The Nature and Possibility of an Environmental Ethic," *Environmental Ethics* 3 (1981): 32.

ues, as dollar amounts reflecting the strength of felt preferences, is confused. The confusion can be illustrated by reference to the example of the teenager who received a symphony ticket instead of the coveted rock concert ticket.[37] In the example, the teenager tried to give away the symphony ticket, thereby demonstrating that it had no demand value for her before the concert. She had no felt preference to own or use the ticket; its value was in transforming her future felt preferences, not in fulfilling her present ones. Transformative values do not exist in the fixed realm of unquestioned felt preferences and demand values. It is true that when the grandparents bought the symphony tickets, they registered a demand for them and, by extension, a demand value associated with the granddaughter experiencing a live performance. But this results from the particular facts of the hypothetical case. Had the grandparents purchased some other gift, the symphony tickets would nevertheless have had (unrecognized and unregistered) transformative value for the granddaughter.

When species preservationists value species for their transformative value, they hope that, given enough exposure to wild species and pristine ecosystems, people will change their felt preferences through a complex process of developing a new world view and adjusting preferences accordingly. The value of such changes need not, logically, be expressed on the scale that is fixed for the purposes of a BCA. Analysts must accept stated, felt preferences as givens; otherwise, they cannot express preferences on their single, monetary scale of values. But transformative values refer to alterations, through time, of that temporarily frozen scale and therefore may not be expressed on that scale.

This chapter addressed three concerns that lead me to expect that economic analyses will systematically understate the true benefits of species.

First, the framing of the question of the value of particular species (or of aggregated values of collections of species) on a BCA approach provides no place for the presumption that every spe-

---

[37] See Section 1.2.

cies has prima facie value as a unit of total diversity. This value was only sketched here and will be developed in much more detail in the next two chapters.

Second, I claimed that many possible and future benefits derived from species cannot be given dollar figures, and therefore that the first assumption of the BCA approach is false. The next two chapters will be devoted to a discussion of ecosystem services and the difficulties in recognizing, let alone quantifying, the full contribution of species to ecosystems.

Finally the BCA approach must assume that all values are reducible to the value of preference satisfactions. In this way they can be given a market value. This amounts, in my terminology, to claiming that all values can be expressed without distortion as demand values. But neither intrinsic values nor transformative values can be expressed accurately as demand values. Accordingly, they will not be discussed further in this part. Part B will be devoted to a discussion of claims that nonhuman species have intrinsic value. Part C will explain and defend the view that nonhuman species have transformative value.

T H R E E

THE VALUE OF ECOSYSTEMS AND THE VALUE OF SPECIES

3.1 Ecosystem Services

"Man has generally been preoccupied with obtaining as much 'production' from the landscape as possible by developing and maintaining . . . ecosystems [such as] monocultures. But, of course, man does not live by food and fiber alone," says noted ecologist Eugene Odum. He goes on to cite the need for a balanced CO_2-O_2 atmosphere, for the climatic buffer provided by oceans and masses of vegetation, and for clean water. These, Odum notes, are provided by the "less productive" landscapes. "In other words," he concludes, "the landscape is not just a supply depot but is also the *oikos*—the home—in which we must live."[1] Since ecosystems are composed of plants and animals, individual species are, indirectly, the true source of these benefits. "Ecosystem services," however, are seldom included in benefit—cost estimates of the value of species. This is no accident; they are not priced in markets, and it is difficult even to conceive a dollar value for services such as oxygen production. Imagine asking someone on a questionnaire how much payment they would require to accept willingly an oxygen-poor atmosphere. An accurate assignment of dollar values to species would somehow have to attribute values to each ecosystem service and then apportion these values to all the particular species that contribute to their production. This task has never been completed for even one

[1] Eugene P. Odum, "The Strategy of Ecosystem Development," *Science* 164 (1969): 266.

service or one species. The difficulties of computation are legion and derive from many sources.

In Section 3.2, I will show that the project of assigning values to the contributions of species through ecosystem functioning to present economic and social activities is beset with problems on two levels. First, attempts to assign economic values to functioning systems, such as attempts to ecologists of the early 1970s to assign dollar values to tidal wetlands, have proved highly controversial. But, second, these problems appear minor compared to those that would inevitably arise if one were to try to factor those ecosystem values into values assigned to individual species.

There is also the danger of ignoring certain services, because most ecosystem services are easily taken for granted. This is especially a problem regarding future values to be derived from ecosystems. Since the loss of species is irreversible, values of ecosystem services should be projected into the indefinite future, while the economic activities that damage systems and extinguish species normally contribute to human value over a limited period of time. This difficulty will be dramatized in Section 3.3, where the creative, contributory value of species, the ability of species to generate new systems and repair damaged ones, will be highlighted. The argument that diversity augments diversity—and that losses in diversity create a cascade of further losses—establishes the contributory value of species.

In Section 3.4 the contributory value of species is marshaled to produce two arguments that the loss of any species is a cause for considerable concern. In Section 3.5, a third argument is given, that each extinction runs a small risk of ecological catastrophe. In that section it will also be shown that these contributions of species to systems, to services derived from systems, and to the creativity of nature, while of indisputable importance, cannot be assigned dollar values.

3.2 Present Services

As policy makers in the late sixties and early seventies perceived the ever-increasing infringement of human development on ecosystems, they turned to ecologists for guidance. Recognizing that losses of ecosystem services are distributed across the population and that they are "external" to economic markets, these policy makers asked, for example, about the value of an acre of undeveloped land or estuary. Ecologists accepted this challenge by suggesting dollar figures representing the value of ecosystem services, both commercial (as in contributions to fisheries) and noncommercial (as in tertiary waste treatment). The most celebrated example of this exercise was the attempt to assign a dollar value to an acre of tidal wetlands. A brief examination of that attempt will be instructive here.

In 1962, John Teal published a paper synthesizing studies of production, respiration, and animal abundances in a salt marsh ecosystem in Georgia.[2] That paper ended with the conclusion that "the tides remove 45% of the production before the marsh consumers have a chance to use it and in so doing permit the estuaries to support an abundance of animals."[3] This observation provided an impetus toward the quantification of the value of salt marshes as the basic source of commerical fisheries. Following this lead, James G. Gosselink et al. published a 1974 report, *The Value of the Tidal Marsh*, in which an acre of salt marsh was estimated to have a value of approximately $2,500 per year for the tertiary treatment of chemical nutrients.[4] As Scott Nixon points out, however, this figure was not supported by nutrient uptake studies and, worse, the claim of "outwelling" (the exportation of nutrients to the ocean) seems to contradict the claim that salt marshes act as nutrient sinks.[5] While the figures derived from these studies were often used in fighting proposed wetlands de-

[2] J. M. Teal, "Energy Flow in the Salt Marshes of Georgia," *Ecology* 43 (1962): 614-624. Also see Scott Nixon, "Between Coastal Marshes and Coastal Waters— A Review of Twenty Years of Speculation and Research on the Role of Salt Marshes in Estuarine Productivity and Water Chemistry," in *Estuarine and Wetland Processes with Emphasis on Modelling*, ed. Peter Hamilton and Keith B. MacDonald (New York: Plenum Press, 1980), pp. 437-525.
[3] Teal, "Energy Flow in the Salt Marshes," p. 624.
[4] James G. Gosselink, Eugene P. Odum, and R. M. Pope, *The Value of the Tidal Marsh* (Baton Rouge, La.: Center for Wetland Resources, 1974). Quoted in Nixon, "Between Coastal Marshes and Coastal Waters," p. 443.
[5] Nixon, "Between Coastal Marshes and Coastal Waters."

velopment, the concepts central to them have resisted empirical documentation, and there are strong reasons to doubt the figures provided. Nixon concludes that the data now available do not support the view that marshes are important factors in coastal marine nutrient cycles, but he emphasizes how tentative the data are concerning these matters.[6]

This example illustrates the extraordinary difficulty involved in assigning dollar values to natural systems. Data gathered over a twenty-year period of intense research on detailed questions concerning just two possible ecosystem services proved inconclusive. Under the circumstances the quantitative measures suggested are nearly worthless. But salt marshes are among the most simple of stable ecological systems. Ecologists are even further from understanding more complex systems and now doubt, in general, their ability to offer monetary descriptions of the value of ecosystems or of the services they provide. The intent of emphasizing this point here is not to suggest, of course, that ecosystems do not provide important services but rather that the services offered are so diffuse and untraceable as to defy even general estimates of value.

If the literature on the value of ecosystems has proven inconclusive, that is only the beginning of problems for those who hope to assign dollar values, indirectly, to the services derived from species through their contribution to ecosystem services. Assume, as I have just argued is not the case, that there exists a set of reliable figures describing the dollar values of all the services derived from a particular ecosystem. To be useful in assigning a dollar value to individual species, these figures would have to be apportioned among the various species in the system. Surely it would be inaccurate merely to divide the amount by the number of species because this would falsely assume that all species contribute equally to providing the services. An accurate assignment would require detailed knowledge of nutrient cycles, drainage patterns, the contributions of particular species to them, et cetera.

But even this unavailable information would be insufficient to assign fair portions of the value to species, as some species might

[6] Ibid., p. 509.

do nothing, in themselves, to provide a service and yet might be necessary for the continued existence of another species that is essential to the production of the service. For example, suppose that a particular flowering plant is essential to moderating the hydrological cycles of an area. How should the value of that service be apportioned between the flowering plant and the insect species that pollinate the plant? To answer this kind of question one would have to know the detailed interrelationships and interdependencies among all the species. This requires, in turn, knowing which species are "keystone" species, knowing which species are redundant, and knowing how much the life cycles of various species can be disturbed before they can no longer support other species dependent upon them. Scientists are very far from having this sort of detailed understanding of any particular natural ecosystem.

For all these reasons dollar figures assigned to the value of species' contributions to ecosystem services would be essentially arbitrary. Such figures would depend on so many unjustified assumptions about ecosystem functioning that no credence could be given them. And yet economic analyses of the value of any species would, by common consent, be radically incomplete if they ignored that species' contribution to commercial and noncommercial services derived from ecosystems.

3.3 Self-Augmenting Diversity Spirals

I have just argued that we lack the scientific knowledge to make accurate assessments of the benefits currently derived from ecosystem services and, especially, to apportion such benefits to individual species. Assessing these benefits as they accrue into the indefinite future is even more difficult. Species adapt and evolve in response to the abiotic environment and to the other species that form their biotic environment. As systems become more complex and interconnected, new symbioses, predation relationships, and interdependencies of all sorts develop. Each of these interdependencies represents a solution to an environmental problem for an organism and, simultaneously, a new opportunity for other species to find a niche within the developing system. Over evolutionary time such relationships promote adaptation and creation of new species. I refer to the role of the species in producing new adaptations, new relationships, new niches, and new species as the contributory value of species. Species contribute to the long-term creativity of nature—an area of great biotic diversity creates more diversity through time. I now turn to a presentation and explanation of the empirical evidence that diversity is self-augmenting. Since ecological processes contribute to evolutionary ones, I begin with a description of how diversity creates diversity in ecological time.

A. Ecological Time

As was noted in Section 2.2, it is important to distinguish between three kinds of diversity: within-habitat, between-habitat, and total. The concern of this book is with the preservation of species and, as was explained in Section 2.2, total diversity over geographic areas is therefore the crucial focus. Species preservationists need not be concerned with the extirpation of a species from particular ecosystems when these result from natural successional developments. Consequently, the present concern is with how total diversity protects and encourages total diversity. However, since the total diversity of an area is directly related to its within-habitat and between-habitat diversity, all three concepts are relevant.

Classical theories of succession posited sharply differentiated developmental stages. For each combination of abiotic conditions (climate, soil type, et cetera) it was believed that an internal dynamic made inevitable a succession toward a single, stable association of species called a "monoclimax." Theories now in favor minimize attribution of vegetational patterns to organic, community-based causes and tend not to predict sharp boundaries between stages or a culmination in a stable monoclimax.[7] The alteration of community structure is attributed rather to differential

[7] See, for example, William Drury and Ian C. T. Nisbet, "Succession," *Journal of the Arnold Arboretum* 54 (1973): 331-368.

growth and dispersal rates of particular plant species, and the distribution of animal species will be to a considerable extent determined by these structures. The opportunist species that are well suited to colonize spaces opened by disturbances are usually not good competitors in the long run. So, it can be expected that species with fast growth rates, short life spans, and high dispersal rates will provide the dominant structure immediately after a disturbance and that long-lived plants which divert energy from propagation to the development of biomass will predominate later. The picture that emerges from this modified theory of succession is that of a patchy landscape. Because disturbances occur at irregular intervals and affect areas of varied sizes, and because recolonization will be affected by random factors of dispersal, the result will be a harlequin environment varying in species makeup across space and time.[8] The vegetational structure of the community will, in turn, determine which animal species are likely to exist and dominate in various patches.

Species colonize disturbed areas and compete for niche space with species already present and with other invaders. Consequently, one finds a mosaic of habitats and microhabitats. Total diversity in such areas is likely to be quite high,[9] even though species may be subject to frequent local extirpations. Simon Levin described these processes illuminatingly:

[M]ovement between patches is generally thought of as having a homogenizing effect, but this need not be the case. Fu-

[8] For classic accounts of patchiness in succession, see A. S. Watt, "Patterns and Process in the Plant Community," and R. H. Whittaker, "A Consideration of Climax Theory: The Climax as Population and Pattern," both in *Ecological Monographs* 23 (1953). Comprehensive recent accounts include John L. Harper. *Population Biology of Plants* (London: Academic Press, 1977), Chapter 23; Peter S. White, "Pattern, Process and Natural Disturbance in Vegetation," *The Botanical Review* 45 (1979): 229-297; Simon Levin, "Dispersion and Population Interactions," *American Naturalist* 108 (1974): 207-228; and Levin, "Population Dynamic Models in Heterogeneous Environments," *Annual Review of Ecology and Systematics* 7 (1976): 287-310. See also Drury and Nisbet's positive account of spatial and temporal sequences of vegetational patterns in "Succession."

[9] Harper, *Population Biology*, p. 711; E. C. Pielou, "Species Diversity and Pattern—Diversity in the Study of Ecological Succession," *Journal of Theoretical Biology* 10 (1966): 370-383.

gitive strategies maintain species in the community through spatio-temporal patterns which may initiate, aggravate, and capitalize upon phase differences. The fugitive is doomed locally, but survives globally by a balance between dispersal ability and competitive or escape ability. It is only a slight extension of this notion to replace direct dispersal from patch to patch by indirect (and delayed) dispersal. . . . When this generalization is made, the fugitive becomes the rule rather than the exception, since most species are, to some degree, locally ephemeral.[10]

The total diversity of an area is constituted by, and at the same time maintains and increases, this patterned patchiness or mosaic of community structure because the availability of seeds, propagules, and migrating animals is an essential part of ecosystem de velopment.

It is impossible to overemphasize the importance for the development of community structure of a large and diverse pool of species available to colonize open spaces. One evidence of this importance derives from differences between primary and secondary succession. Informally, a secondary succession follows a disturbance such as a fire, which destroys vegetation but leaves seeds and perhaps some living roots to begin the process of renewal. Primary successions follow rare events, such as volcanic eruptions, exposure of sand banks, receding glaciers, and mine tailings,[11] which leave a highly altered soil lacking seeds and propagules, as well as vegetation. While these phenomena have not been carefully compared, it is widely agreed that primary succession proceeds much more slowly than does secondary succession.[12] Further, the larger the area disturbed, the slower primary succession will begin and progress.[13] These facts suggest that im-

[10] Simon Levin, "Population Dynamics and Heterogeneous Environments," *Annual Review of Ecology and Systematics* 7 (1976): 294.

[11] Drury and Nisbet, "Succession," pp. 357-358.

[12] Ibid., p. 359.

[13] G. E. Likens, F. H. Bormann, R. S. Pierce, and W. A. Reiners, "Recovery of a Deforested Ecosystem," *Science* 199 (1978): 495.

portant agents governing the rate of successional development are the availability and proximity of colonizing species.

This generalization holds for later stages of succession as well. Martin Cody, noticing great variation in between-habitat diversity in cross-continent comparisons of otherwise similar areas, speculates that an important factor is "the relative accessibility of each habitat to a species pool of more or less appropriate colonists from elsewhere on the continent."[14] Cody found that approximately 50 percent of the variation in between-habitat diversity was explained by differences in colonist accessibility.[15] This account squares with the data from the studies of island biogeography carried out by Robert MacArthur and E. O. Wilson.[16]

The importance of proximate colonizers in the development of community structure is, then, well established. It will be instructive, however, to explore in more detail the means by which the total diversity of an area promotes within-habitat diversity and, in turn, between-habitat diversity. Robert and John MacArthur discovered that the number of bird species present in a habitat was predictable from its foliage height diversity.[17] That is, observations of the density of foliage at varied points from the ground provided a profile indicating the diversity of habitats available, regardless of the number of plant species present.[18] This surprising discovery led Robert MacArthur to speculate that within-habitat diversity is largely determined by structural features of the habitat and, consequently, that there must be a fairly uniform level at which within-habitat diversity of bird species is maximal and not likely to develop further.[19] Since there seems not to be

[14] Martin L. Cody, "Towards a Theory of Continental Species Diversity," in *Ecology and Evolution of Communities*, ed. Martin L. Cody and Jared M. Diamond (Cambridge, Mass.: Harvard University Press, 1975).

[15] Ibid., p. 244.

[16] Robert H. MacArthur and Edward O. Wilson, *The Theory of Island Biogeography* (Princeton, N.J.: Princeton University Press, 1967), pp. 23, 171, 176; also see Robert K. Peet, "Ecosystem Convergence," *American Naturalist* 112 (1978): 441-444.

[17] Robert MacArthur and John MacArthur, "On Bird Species Diversity," *Ecology* 42 (1961): 594-598.

[18] Ibid., p. 597.

[19] Robert H. MacArthur, "Patterns of Species Diversity," *Biological Review* 40 (1965): 510-533.

such a uniformity in areawide species diversity, he concluded that "during the initial stages of colonization the diversity of species will be wholly within-habitat; and eventually all new diversity will be between-habitat."[20] Available evidence suggests that greater total diversity of bird species, such as that exhibited in the tropics over that in the temperate zones, is due more to greater between-habitat diversity than to greater within-habitat diversity.[21]

Without questioning MacArthur's conclusion that bird habitats reach species saturation in ecological time, R. H. Whittaker has argued cogently that no such limits are operative in vegetational diversity.[22] In either case the availability of colonizers affects total diversity, and this is the main point. Where saturation limits apply, intense competition for available niches accelerates habitat differentiation and, in turn, the development of new types of habitats.[23] Where no such limits apply, within-habitat diversity increases indefinitely and also can, in the face of environmental (biotic and abiotic) variation, create differences in local structure that create new habitats. In either case total diversity is enhanced because it is a product of within-habitat diversity and between-habitat diversity.

The mechanisms that increase diversity (in the absence of catastrophes) are too complex to be described in detail here but several are worth mentioning briefly. As an area recovers from a severe disturbance, the species that are dominant in the earlier stages are the fast colonizers with high dispersion rates, fast growth, and short life spans because they are suited to occupying open spaces. Consequently they are weaker in competitive ability

[20] Ibid., p. 523.

[21] Ibid., p. 529.

[22] R. H. Whittaker, "Evolution of Diversity in Plant Communities," in *Diversity and Stability in Ecological Systems*, ed. G. M. Woodwell and H. H. Smith, in *Brookhaven National Laboratory Publication No. 22* (Springfield, Va.: Clearinghouse for Federal Scientific and Technical Information, 1969), pp. 178-196, esp. p. 188; R. H. Whittaker, "Evolution and Measurement of Species Diversity," *Taxon* 21 (1972): 213-251, esp. pp. 241-244. Also see Harper, *Population Biology*, pp. 712-713.

[23] Some of the mechanisms by which this process occurs will be described in the next two paragraphs.

than are species that put less energy into dispersal. As the open space is filled, plant species adapted to longer life spans and to the development of biomass begin to dominate by shading other species.[24] As time passes and more species attempt to invade and hold territory, selective pressures will confine populations to the microhabitats to which they are best suited, resulting in greater patchiness.[25] In the process the niche breadths of the species present (either established or invading) will be narrowed, and more species will be "packed" along the resource gradient.[26]

As competition stiffens, species with mechanisms designed to exclude competitors will gain advantages. Besides such obvious mechanisms as exclusion by occupation of space and deprivation of nutrient sources (for example, through shading), ecologists are giving increased attention to chemical mechanisms.[27] For example, plant species exclude competitors on the same trophic level by placing inhibiting chemicals in the soil. These inhibitors further differentiate microhabitats.

Other mechanisms contributing to species diversity include methods whereby species inhibit their own propagation. Surprisingly, certain plants introduce into the soil around them chemicals inhibiting the growth of their own seedlings.[28] The seedlings of some forest tree species are more likely to reach the sapling stage if they root under the canopy of a tree of another species.[29] These chemical adaptations often work between trophic levels as well. Plants develop chemical structures that taste bad or even poison herbivores. But these same mechanisms can serve as signals to other species, for example, to insect pollinators.[30] Whenever a plant develops a special chemical defense, it opens possi-

[24] Harper, *Population Biology*, p. 712.
[25] Whittaker, "Evolution and Measurement," p. 216.
[26] MacArthur, "Patterns," pp. 522-523.
[27] For a review, see R. H. Whittaker and P. P. Feeney, "Allelochemics: Chemical Interactions between Species," *Science* 171 (1971): 757-771.
[28] Robert MacArthur, "Species Packing and Competitive Equilibrium for Many Species," *Theoretical Population Biology* 1 (1970): 1-11.
[29] J. P. Grime, *Plant Strategies and Vegetation Processes* (New York: John Wiley and Sons, 1979), pp. 180-181.
[30] Whittaker and Feeney, "Allelochemics"; Paul R. Ehrlich and Peter H. Raven, "Butterflies and Plants: A Study in Coevolution," *Evolution* 18 (1964): 586-608.

bilities to any herbivorous species that, if it can develop a means to evade the defense, will gain an uncontested food supply.

Many other mechanisms also operate across trophic levels to maintain diversity.[31] Grazing by herbivores increases plant diversity in two ways: first, the pressure of consumption prevents overdominance of any one species of plant; and, second, the herbivores exert different control mechanisms and thereby increase niche possibilities on the lower level.[32] The extraordinary diversity of forest tree species in the tropics apparently results at least in part from the pressures of predation on seeds and seedlings. Concentrations of a particular species leave seeds vulnerable to concentrations of seed-consumers drawn by the abundance of food. So, it is a necessity to distribute seeds widely. The result is a highly mixed and diverse collection of canopy species.[33]

All these mechanisms contribute to greater total diversity. Species with specialized adaptations have advantages in competition and, at the same time, create opportunities for additional colonizers. Greater total diversity expresses itself in a more heterogeneous environment. This heterogeneous environment supports—indeed, constitutes—total diversity, the result of high degrees of within- and between-habitat diversity. Such a totally diverse area is also the source of colonizers that intensify competition for niche space in habitats, force habitat differentiation, and create still more heterogeneity. These processes explain why total diversity is self-augmenting in ecological time.

B. Evolutionary Time

The same processes explain, indirectly, how diversity is also self-augmenting in evolutionary time. Whittaker describes the general process as one of creating new niche spaces along "axes." An axis is one particular determinant of survivability within a niche, such

[31] Harper, *Population Biology*, pp. 735-736.

[32] J. L. Harper, "The Role of Predation in Vegetational Diversity," in *Diversity and Stability in Ecological Systems*, pp. 48-62; Whittaker, "Evolution and Measurement," p. 216.

[33] Daniel H. Janzen, "Herbivores and the Number of Tree Species in Tropical Forests," *American Naturalist* 104 (1970): 501-528.

DEMAND VALUES

as a source of food, availability of shelter from predators, et cetera. The number of spaces increase in evolutionary time "as additional species enter the community, fit themselves in between other species along the axis, and increase the packing of species along axes."[34] Species can be added at the "ends" of axes as well, as some species specialize in the use of marginal resources. Some species create entirely new resource options. As an example of this, Whittaker cites the evolution of modern flowering plants and the resulting evolution of bird species that rely on nectar and fruits of these plants. Diversity thus increases through evolutionary time by creation of new axes, by lengthening axes, and by packing more and more species into narrower niches on the various axes.

Interacting groups of species, such as plants and animals, predators and prey, and between plants and symbiotic fungi, contribute to increasing diversity in two basic ways. First, each species provides resources for others to use. The greater the diversity of plant species, for example, the greater the variety for grazing species. Second, these influences work in the opposite direction as well because grazing animals may also increase diversity among plants by preventing any plant species from becoming too strongly dominant. Whittaker concludes: "We can thus say that diversity begets diversity. Species diversity is a self-augmenting evolutionary phenomenon; evolution of diversity makes possible further evolution of diversity."[35]

It is important to note that in Whittaker's description the entire process gains its initial and ongoing impetus "as additional species enter the community."[36] The existence of areawide, total diversity is, then, a prerequisite for this process, and greater degrees of total diversity will intensify and accelerate the process.

It is again useful to distinguish the means by which within-habitat diversity is generated from those means that generate between-habitat diversity through considerable evolutionary time.

[34] R. H. Whittaker, *Communities and Ecosystems* (New York: MacMillan, 1970), p. 103.
[35] Ibid.
[36] Whittaker, "Evolution and Measurement," p. 213.

Does this process achieve a point of saturation—a point at which diversity is maximal and unable to develop further? Apparently not, at least not where the environment is sufficiently stable so that species tend to evolve mainly in response to biotic factors such as interspecific competition. Where climate and other abiotic factors largely determine survival, saturation may be reached, as the number of genetic lines that can adapt to the environmental conditions may be limited. There seems also to be one further qualification. Within-habitat diversity may reach a saturation point for some taxonomic groups whose niches are largely determined by food sources and feeding strategies (as MacArthur pointed out for bird species[37]). But since diversity can still increase by habitat differentiation and since other taxonomic categories, especially vascular plants, do not achieve within-habitat saturation, there seems to be no upper limit on either within-habitat diversity or between-habitat diversity. And since total diversity is a product of those forms of diversity, it, too, is apparently unlimited.

C. Diminutions in Diversity

This spiral works in reverse as well. Losses in diversity beget further losses, and the upward diversity spiral will be slowed and eventually reversed if natural and/or human-caused disturbances are severe and prolonged.

If a species goes extinct, other species that interact with it and depend upon it are threatened. A well-known biologist has asserted that for every plant species that becomes extinct fifteen animal species can be expected to follow.[38] In general, species that have evolved exclusive relationships, feeding or otherwise, with other species are more likely to suffer extinction because their in-

---

[37] MacArthur, "Patterns."

[38] Peter Raven, Statement at Hearings before the Subcommittee on Environmental Pollution of the Committee on Environment and Public Works, United States Senate, 97th Congress, December 8 and 10, 1981 (Washington: U.S. Government Printing Office, 1982), p. 293; also see John Terborgh and Blair Winter, "Some Causes of Extinction," in *Conservation Biology: An Evolutionary Ecological Perspective*, ed. M. E. Soule and B. A. Wilcox (Sunderland, Mass.: Sinauer Assoc., 1980).

herent probability of extinction is increased by the likelihood of the species on which they depend becoming extinct.[39] Even species that have no exclusive dependencies face a considerable risk deriving from the importance of any relationships they do have. Systems that are characterized by high degrees of complexity and interdependence among species and by large numbers of specialized species—the ones with the greatest amount of biological diversity—are precisely those that suffer most seriously from disturbances that extirpate one or more species from them. Thus, more complex systems are in some ways less stable than simple ones.[40] Also, the simplification of systems relaxes the pressures of natural selection on remaining species, lessening the need for adaptive specializations, so the reversal of the diversity spiral operates in evolutionary as well as ecological time. Thus, the contrapositive of the ecological law that diversity begets diversity is likewise true.

Total diversity counts travel in spirals: either total diversity increases in an accelerating spiral or it decreases in an accelerating spiral. If disturbances sufficient to extinguish species interrupt the natural tendency toward the generation of more diversity, the upward spiral will be slowed. If those disturbances continue unabated and more extinctions occur, then the spiral can be reversed and an accelerating downward spiral can begin. These two related ecological generalizations provide the basis for several powerful arguments for species preservation.

3.4 The Contributory Value of Species

A. Argument 1

That diversity produces diversity does not, of course, imply that diversity is valuable. To show that, one must show that diversity

---

[39] Charles W. Fowler and James A. MacMahon, "Selective Extinction and Speciation: Their Influence on the Structure and Functioning of Communities and Ecosystems," *American Naturalist* 119 (1982): 483; Terborgh and Winter, "Some Causes of Extinction."

[40] These points will be developed more fully in Chapter 4, where I will discuss the diversity-stability hypothesis at length.

is preferable to simplicity—the central issue in species preservation. Here the familiar species-by-species accounts of the value of species become relevant. If important values are known to be served by some identified species and if it is also assumed that other, as yet unidentified and/or unexamined, species will prove useful in the future, then increments in diversity increase the likelihood of utilitarian benefits to humans. The strategy of this discussion, outlined earlier, now bears fruit. If diversity contributes to diversity, and if it can be assumed that some significant subset of any random collection of species will prove useful to humans, then there is reason to believe that any given species will be of some value, direct or indirect. Even species that have been carefully, but unsuccessfully, examined for direct human uses may still be useful because they contribute to increases in diversity that in turn contribute to the generation of more species. Of those species, some will predictably prove useful to humans, and the species that were of no direct use may be useful indirectly in supporting the ecosystems on which useful species depend.

Even species that fail in the competition for niche space in a particular ecosystem have made an important contribution to the structure of that system. By intensifying the competition for niche space, they have, even in the process of being extirpated by superior competitors, exercised adaptational pressure on those species, caused further specializations, and developed more tightly packed and more highly integrated species relationships. Consequently, even the losers in such competitions are important. The ecosystem development and total diversity of an area are served, both directly and indirectly, by the existence of every species. That is, even if a species ultimately disappears because of its inability to compete, it still may have contributory value.

To this line of reasoning it might be objected that while certain minimal levels of diversity are highly valuable, the loss of a single species from an already diverse system is of negligible importance. This objection might be strengthened if it were also noted that many of the species that are endangered are naturally rare and have little apparent effect on other species.

This objection ignores three important facts. First, while complex systems have considerable redundancy, it is precisely this re-dundancy that creates and sustains the climate of competition driving the processes just described. Even if a species is rare in an ecosystem, it may be rare because it is competitive in a very nar-row niche space, *given the other species in the system*. Extirpation of such a rare species lessens the competitive pressure on the dom-inant species. Second, we usually have insufficient knowledge of interdependence to judge which species are, in fact, redundant. Even very rare species can exert considerable influence on an eco-system. Third, because species are involved in a complex web of interdependencies, the loss of one species from this web may re-sult in a gradual decline in other species, eventually pushing the system below a threshold beyond which further declines are in-evitable.

B. Argument 2

Given that diminutions in diversity beget further diminutions, a second argument for protecting species emerges. Scientific under-standing of ecosystems is too limited even to begin to list inter-dependencies among species, so it is impossible to predict which species will be included in the cascading wave of extinctions re-sulting from an initial extinction. When an extinction creates more extinctions, a downward spiral in diversity, which will be extremely difficult to reverse, is begun. Any decision to extinguish a species or to allow one to go extinct is thus one step toward a policy that places higher priority on economic development than on protecting biological diversity. The decision to adopt such a policy must be understood as a decision to begin or accelerate a spiralling course toward a less complex and diverse biological world.

Besides obligate mutualisms (that is, complete dependencies of one species on another) there are many lesser dependencies as well. These are often very subtle, and loss of one species can cause small changes in the habitats of other species, making their envi-ronment slightly less suitable and beginning a slow deterioration in their population. For example, one species may have its popu-lation reduced because the extirpation of another lessens but does not eliminate its food supply. But a third species may have a stronger dependency upon the diminished species so that it cannot be supported at all by the smaller population. Or, the diminished population might be lost years later because it has become more susceptible to random environmental fluctuations. The full effects of extinctions, especially multiple ones, may continue for centuries.[41] Ecologists cannot delineate all these relationships and cannot provide economists with the information necessary to compute the economic effects of a species loss or the consequences in further losses. Therefore, a computation of the true value of a species to the human race, if projected over unlimited time, must include all the unpredictable ramifications of its extinction—a task to which neither ecologists nor economists are equal. But it is possible to infer that the true value is sometimes extremely large. A decision to limit value computations regarding losses in diversity to brief time periods, such as thirty years, is to sacrifice the future options of the human race for short-term economic gain.

Because species exist in ecosystems too complex for current human understanding, it is a mistake to conceptualize a loss of a species as simply that: it is more accurately viewed as a first step in a process of ecosystem simplification. Viewed in this manner, it is more likely that the gravity of the losses will be recognized, even if it is impossible to place dollar figures on them.

The first two arguments, deriving from the law that diversity begets diversity and its contrapositive, establish that the prima facie value of a species is significant. When higher-order dependencies of other species on extinguished ones are taken into account, it becomes likely that some species eventually lost in the processes that may be begun by the extinction of just one species would have proved to be of substantial worth to humans.

[41] For example, the most recent theories describing events occurring at the time of mass extinctions at the end of the cretaceous era suggest that a single major event (most likely the striking of earth by an asteroid or disintegrated comet) led to a cycle of extinctions that occurred over a period of fifty thousand years. See Kenneth J. Hsu et al., "Mass Mortality and Its Environmental and Evolutionary Consequences," *Science* 216 (1982): 249-250.

3.5 Extinctions and Zero-Infinity Dilemmas

Some skeptic might respond that, while some useful species are almost certainly lost as a direct or indirect result of an extinction, there are so many species and we have so little research time to examine them that the loss of a few useful ones is not very significant. As long as many species remain, having a few more would be relatively inconsequential.[42] The skeptic might go on to say that protecting other species is a good thing to do, but we have no strong obligation to protect them. In general, ethical theorists believe that the obligation to be beneficent (as in giving to the poor) is weaker than the obligation not to harm someone.[43] This reasoning might be applied in the present case. In a situation where species are abundant, the protection of a handful is only an act of beneficence—providing more of a good thing to future individuals who will already be well blessed. This weaker obligation might be thought to be easily overridden by matters of convenience and short-term economic gain. That a great deal of diversity is essential, in the long run, for human health and well-being does not, on this view, entail that diversity must be maximized. The obligation to protect a considerable amount of diversity is much stronger, the argument concludes, than the obligation to protect each and every species.

A first response to this skeptical objection is that it seems insensitive to the cascading effect of species losses—one is virtually never discussing the loss of just one species. The loss of a species represents the beginning of a trend. This initial response can be built into a full-fledged argument.

Note, first, that the skeptical objection carries different weight, depending on how one views the present situation. If biological diversity on the planet is very great and accelerating, then the loss of a species merely slows the acceleration of the upward spiral. It would then be plausible to consider the obligation to preserve a

---

[42] This objection was suggested to me by Henry Shue and Robert Fullinwider.

[43] See, for example, William K. Frankena, *Ethics* (Englewood Cliffs, N.J.: Prentice-Hall, Inc., 1963), pp. 36-37.

species as an obligation merely of beneficence rather than as a stronger obligation to avoid causing harm. The skeptical objection rests on an empirical assertion that the biological world is, in fact, still increasing in diversity. But this empirical assertion is demonstrably false. The current rate of extinctions is already far in excess of the historical rate of speciation, and it is increasing by leaps and bounds.[44] So, the loss of a species is not merely a slowing of the spiral, as it would have been a few centuries ago.

Further, the accelerating nature of the downward spiral implies that each new species loss is more important than the one preceding it. Norman Myers estimates that extinction rates are accelerating as follows. Until the appearance of man, there was 1 extinction per 1,000 years, even during the period in which the dinosaurs were eliminated.[45] Since man has developed more advanced technologies that exert a greater force on extinction rates, these rates have increased dramatically over the past 50,000 years.[46] Anthropogenic species extinctions probably reached a total of about 75 from A.D. 1600 to A.D. 1900; about the same number have been eliminated since 1900.[47] The rate continues to increase alarmingly. For much of this century the rate was one human-caused extinction per year, but it has skyrocketed since 1960 and largely because of the progressive destruction of tropical forests as many as 1,000 species are now estimated to be lost each year.[48] If, as many experts have predicted, 1 million species will be lost in the final quarter of the present century, this would represent an average of 40,000 lost per year.[49] By the end of the century one species will be lost each hour.[50] While these figures are not based upon precise data, there is no question that the rate is

[44] Paul Ehrlich and Anne Ehrlich, *Extinction: The Causes and Consequences of the Disappearance of Species* (New York: Random House, 1981), p. 7; Thomas Eisner et al., "Conservation of Tropical Forests," *Science* 213 (1981): 1,314.

[45] Norman Myers, *The Sinking Ark* (Oxford: Pergamon Press, 1979), p. 4.

[46] Ibid.

[47] Ibid. Myers acknowledges that these figures are artificially low, as they are limited almost entirely to birds and mammals and do not count the losses of unknown and unnamed species (p. 31).

[48] Ibid., p. 4.

[49] Ibid., p. 5.

[50] Ibid.

accelerating and that any rate approximating the projected range would be unprecedented in the recorded history of the planet (including the evidence of the geological record).[51] Speciation rates vary across species groups but are certainly much lower than current and projected extinction rates.[52] A downward spiral in biological diversity has clearly begun.

On an area-by-area basis the situation is even more grave than it is in terms of worldwide biological diversity. As frightening as are current and projected rates of worldwide extinctions, the rate of local habitat destruction and ecosystem simplification are even more so. Local extirpations result from human population growth and the attendant fragmentation of natural habitats. Biotic communities all over the world are becoming progressively impoverished, even when local extirpations do not involve worldwide extinctions. Reintroductions of species into their historical ranges almost never work and, when they do, are extremely costly.[53] As the twenty-first century is approached, many species that are not actually extinct will have such limited ranges that their susceptibility to extinction will be much greater. And even when strong populations manage to survive in a vastly truncated range, the benefits of living in a more diverse area are lost to all humans existing in the areas where the species has been extirpated. The skeptical objection and the response to it therefore set the stage for another argument for protecting species, an argument that multiplies considerably the already significant prima facie reasons for preservation derived in the first two arguments.

Argument 3

A third argument draws on the empirical generalizations concerning diversity spirals supplemented by the fact that the biological world is currently experiencing a quickly accelerating down-

[51] Ibid.

[52] See Steven Stanley, *Macroevolution: Pattern and Process* (San Francisco: Arthur Freeman and Company, 1979), pp. 106-107, 135-136.

[53] Eugene Morton, "The Realities of Reintroducing Species into the Wild: The Problem of Original Habitat Alteration," In *Animal Extinctions: What Everyone Should Know*, ed. R. J. Hoage (Washington, D.C.: Smithsonian Institution Press, 1985), pp. 147-158.

ward spiral in diversity. Because humans exist at the end of various food chains and depend on other species in many, not always well-understood, ways, they are vulnerable to threats to their own existence. Increases in the global rate of extinctions increase the vulnerability of the human species to extinction.

Paul and Anne Ehrlich preface their book, *Extinction*, with a parable called "The Rivet Poppers." A person enters an airplane for a flight but notices a workman prying rivets out of the wings. Under questioning, the workman explains that the rivets can be sold for two dollars each and that this subsidy keeps the price of flying down. When asked about the safety of the practice, the workman replies that it must be safe as it has been going on for some time and no wings have yet fallen off, even after successive rounds of rivet popping.[54] The analogy should be obvious. Although humans have caused thousands of extinctions, no particular extinction has resulted in a major disaster. So, it can be argued that all the evidence from past cases suggests that no disasters result from such practices. The Ehrlichs' point is that we should be no more reassured by this than the prospective passenger should be reassured by the rivet popper's explanation.

The problem of extinction rates that far surpass speciation rates presents to policy makers a problem that has characteristics associated with other environmental risks with low probabilities and serious consequences, sometimes called "zero-infinity dilemmas." If too many species are lost, by increments, from an ecosystem, an area, or the worldwide biotic community, it is possible that a catastrophic ecosystem breakdown would occur. The risk of any particular extinction having catastrophic effects might be quite low, but the consequences would be extremely serious, whether they occurred over a single system or over the globe as a whole. Assigning economic values to such risks is notoriously difficult.[55]

[54] Ehrlich and Ehrlich, *Extinction*, pp. xi-xiv.
[55] Talbot Page, "Keeping Score: An Actuarial Approach to Zero-Infinity Dilemmas," Social Science Working Paper no. 248 (Pasadena, Calif.: Division of Humanities and Social Sciences, California Institute of Technology, January 1979); also see Talbot Page, "A Generic View of Toxic Chemicals and Similar Risks,"

Actually, the rivet popper's explanation embodies an inductive fallacy—it shows the danger of assuming that the absence of a certain low-probability disaster to date indicates a low likelihood that one will occur in the future. Two kinds of situations must be distinguished. In cases where the individual occurrences of a certain kind of event are independent, for example, nuclear reactor accidents, the absence of such an occurrence can be considered to be (very weak) confirmation of a low-probability assignment to the occurrence. In cases where occurrences are not independent, as when one knows that an urn contains one red ball and one thousand white ones, each failure to draw a red ball increases the probability that the next will be a red one. Assuming, as most ecologists would, that some minimal number of species is necessary to avoid an ecological disaster, species extinctions are dependent occurrences. Each species extinction increases, however slightly, the probability that the next will prove disastrous. Hence, the rivet poppers fallaciously assume that dependent occurrences are independent.

This aside, the rivet-popping analogy and zero-infinity dilemmas have important logical affinities. Because there are so many species, the likelihood that the next extinction will produce a disaster is very low. But a disaster is unlikely only because ecosystems contain considerable redundancy. This redundancy has developed over eons of time by the processes described in the last section. When human disturbances begin to reverse those proc-

---

Ecology Law Quarterly 7, no. 2 (1978): 207-244. Page lists nine characteristics of zero-infinity dilemmas, all of which seem to apply to species extinctions. (1) Ignorance of mechanism: we lack knowledge about how a possible disaster would occur. (2) Potential for catastrophic costs: a disaster could be very costly in economic terms or even in human lives. (3) Relatively modest benefits associated with the risk gamble: the advantages are small and incremental rather than great and sudden. (4) Low subjective probability: the disastrous effect is unlikely. (5) Internal transfer of benefits: the benefits associated with the risk are transferred through markets and reflected in product prices. (6) External transfer of costs: the adverse effects of the risk exist in the environment rather than through the market (they are economic externalities). (7) Collective risk: the risk is faced by many individuals. (8) Latency: there is an extended delay between the action creating the hazard and the manifestation of its effect. (9) Irreversibility: the effect once it occurs is, in a practical sense, irreversible.

esses, each extinction will increase the very small probability that the next extinction will prove disastrous. That this probability is initially very low (a feature shared with other zero-infinity dilemmas) encourages the fallacious inference that the absence of disaster so far is evidence that disaster will never come. With or without this inference, however, a zero-infinity dilemma is faced as any species nears extinction: there is a very small probability that the next species extinction might involve a serious disaster.

What form this disaster might take is difficult to ascertain in advance. I assume that a large number of human casualties and/or a breakdown of the ecological and economic system of a large geographic area would clearly count as a disaster. It may also be the case that no cataclysmic events will occur but that a slow but inexorable process of impoverishment will result in decreased productivity.

A number of ecologists warn that progressive extinction of species involves risks of the sort described by Page. It is worth noting several of these here.

Orie Loucks has argued, using reasoning related to mine, that overmanagement of ecosystems can lead to their failure to rejuvenate themselves. Diversity progressively deteriorates, and the productive potential of entire areas eventually collapses. Natural ecosystems respond to natural disturbances by establishing rhythmic cycles of productivity. Studying community development in Wisconsin forests, Loucks observed repeating "waveform" patterns of development triggered by random perturbations (in this case, forest fires).[56] The changes that take place in succession are a characteristic series of transient phenomena; taken together they make up a stable system capable of repeating itself every time a new perturbation occurs. This allows for the selection of species that fulfill specialized functions at each of the several phases. This wave phenomenon, then, contributes to the evolution and maintenance of diversity. But human control of environments interrupts the natural rhythmic alternation of high

[56] Orie Loucks, "Evolution of Diversity, Efficiency, and Community Stability," *American Zoologist* 10 (1970): 23-24.

productivity phases with high diversity phases. Over time, human management for increased productivity diminishes diversity and, eventually, undermines the process of natural rejuvenation.[57] The small, incremental value of putting more and more land into productive monocultures could lead, after decades or centuries, to a complete collapse of the productive potential of that land.

The general point that diminutions of total diversity may threaten ecosystem breakdown is given further plausibility by the fact that the chemical composition of the oceans appears to be biotically controlled.[58] Most oxygen production originates in plants living in the oceans, and a significant change in these controlling factors could radically alter atmospheric compositions.

The disastrous events that took place in the dust bowl of the 1930s may provide an example of the sort of breakdown here described, as does the progressive desertification of Sub-Saharan Africa. The economic and psychological costs of such breakdowns are sufficiently cataclysmic to compare with nuclear disasters and other zero-infinity dilemmas. Similarly, the loss of species and consequent loss of total diversity can and do have the same effects—but this may not be recognized as it is difficult to correlate effects with any single loss in diversity because the losses are incremental and the effects are delayed and may be triggered by abiotic causes such as several consecutive drought years.

A number of ecologists, then, believe that cumulative species extinctions (or perhaps even one extinction of a crucial species) could have cataclysmic effects. For example, when human hunting and other factors drove alligator populations to dangerously low levels in the 1960s and early 1970s, biologists perceived a general decline in wildlife populations. During droughts and dry spells, alligators dig holes, called "wallows," that collect and hold water. These wallows, when abandoned, provide the only source of water for many other species. It was feared that if the alligator were actually extinguished, a subsequent drought year might destroy many Florida wildlife populations.

[57] Ibid., p. 25.
[58] Albert C. Redfield, "The Biological Control of Chemical Factors in the Environment," *American Scientist* 46 (1957): 205-221.

Cases where decreased diversity or loss of a single species have caused chain reactions may already have occurred, with attention diverted to abiotic factors that would have been absorbed by more diverse areas. I believe these concerns amount to saying that each decrease in total diversity has the characteristics of a zero-infinity dilemma. Some minimum *threshold* for total diversity seems to exist for different areas. If species extinctions share the characteristics of other environmental risk problems, with the additional characteristic of ascending probabilities as more and more extinctions occur, then they must be treated as zero-infinity dilemmas of serious magnitude.

It is difficult to assign values to such risks. Page notes that the Rasmussen report on the risks of a catastrophic core meltdown of a nuclear generator sets the probability of such an event at one in five billion per reactor year. But he also notes that the direct historical record based on actual reactor years with no meltdown is equally compatible with a risk assessment of one in one thousand chances per reactor year.[59] Mishan and Page argue that such assessments cannot be given any quantitative validity. Rather, they suggest that the issue is whether to adopt one of two rules: "Rule A would countenance the initiation or continuance of an economic activity until the evidence that it is harmful or risky has been established beyond reasonable doubt. Rule B, in contrast, would debar the economic activity in question until evidence that it is safe has been established beyond reasonable doubt."[60]

They then argue that, far from allowing an assessment of costs and benefits, such choices reduce to questions concerning the state of the society, the urgency of the need for more goods of the type that incur the risk, and, ultimately, the attitudinal choice of risk-acceptance or risk-aversion.[61]

They conclude, in discussing the analogous case of ozone de-

[59] Page, "Keeping Score," p. 1.

[60] Ezra Mishan and Talbot Page, "The Methodology of Cost Benefit Analysis with Particular Reference to the Ozone Problem," Social Science Working Paper no. 249 (Pasadena, Calif.: Division of Humanities and Social Sciences, California Institute of Technology, January 1979), p. 71.

[61] Ibid., p. 75.

pletion, that "a valid cost-benefit calculation of actions to protect the earth's ozone shield cannot be undertaken in the present state of our ignorance concerning the relevant physical relationships and, therefore, in the present state of our ignorance concerning the nature and magnitude of the risks posed by existing economic activities."[62] The basic problem here is that the feared danger is so great that *if* there is even a very small likelihood of its occurrence, then one should be cautious. But assessment of very low probabilities is difficult because one needs a very long track record upon which to base them. The only way to get direct evidence for such assessments is to incur the risk involved over a very large number of cases—a very dangerous thing to do when the disaster is irreversible. This provides yet another example of where the first assumption of the BCA approach, as it has been defined here, fails. Adequate economic analyses must be sufficiently flexible to include qualitative reasoning of the type described by Mishan and Page.

While the first two arguments, exploiting the contributory value of individual species to the diversity spiral (either upward or downward), establish that each species has some prima facie value, this third argument enhances that value significantly. When the premise that diminutions in diversity create further such diminutions is supplemented by the premise that the downward diversity spiral is already accelerating at an alarming rate, each species takes on an added value. Each species that is lost carries with it the risk of a catastrophic ecosystem breakdown and increases the risk that the next loss will result in such a breakdown. This powerful argument provides a reason for considering the prima facie value of each species to be significant. Further arguments will be explored in the next chapter.

[62] Ibid.

F O U R

DIVERSITY, STABILITY, AND
AUTOGENIC SYSTEMS

4.1 *Diversity, Stability, and the Conservation Ethic*

Over the past few decades, it has been common for conservationists to appeal to what has been called the diversity-stability hypothesis to support the preservation of species and natural diversity. The following passage from Barry Commoner's *The Closing Circle* illustrates the use of this popular argument:

> The amount of stress which an ecosystem can absorb before it is driven to collapse is also a result of its various interconnections and their relative speeds of response. The more complex the ecosystem, the more successfully it can resist a stress ... Like a net, in which each knot is connected to others by several strands, such a fabric can resist collapse better than a simple, unbranched circle of threads—which if cut anywhere breaks down as a whole. Environmental pollution is often a sign that ecological links have been cut and that the ecosystem has been artificially simplified.[1]

In this passage Commoner is applying an interesting and much-discussed version of the diversity-stability hypothesis first given clear statement by Robert MacArthur in 1955.[2] MacArthur proposed measuring the stability of a system by measuring the number of its alternative pathways through which energy can flow. That is, he used more convenient diversity measures as measures

[1] Barry Commoner, *The Closing Circle* (New York: Alfred Knopf, 1972), p. 38.
[2] Robert MacArthur, "Fluctuations of Animal Populations and a Measure of Community Stability," *Ecology* 35 (1955): 534.

of stability by assuming that more diverse systems are more stable. He justified the plausibility of this assumption by explaining that a system with many pathways, representing an abundance of species organized in a complex food web, tends to equilibrate fluctuations in population, as predators will switch from less abundant to more abundant prey species, lowering population levels in the latter and allowing them to increase in the former. This observation became commonplace and led to further theorizing.[3]

MacArthur's discussion set a precedent for interpreting the intuitive idea that diversity and stability are causally related as the claim that diverse and complex systems will be more stable through time. Charles Elton summarized the data supporting the hypothesis by stating six reasons to believe the hypothesis true:

(1) Evidence inferred from mathematical models of populations suggests that models with few species are inherently unstable.[4]

(2) Laboratory experiments confirm mathematical findings that show small mixes of populations to be unstable.[5]

(3) Habitats on small islands, possessing fewer species, are more vulnerable to invasions from other habitats than are those on continents.[6]

(4) The less diverse habitats of cultivated or planted land are also more susceptible to invasions and outbreaks of pest species.[7]

(5) The very diverse tropical forest ecosystems are less susceptible to invasions by pests.[8]

(6) Orchard spraying, which simplifies ecological relationships and upsets the balance between pests and their natural ene-

[3] G. E. Hutchinson, "Homage to Santa Rosalia, or Why Are There So Many Kinds of Animals?" *American Naturalist* 93 (1954): 145-159; and Joseph Connell and E. Orias, "The Ecological Regulation of Species Diversity," *American Naturalist* 98 (1964): 399-414.

[4] Charles S. Elton, *The Ecology of Invasions by Animals and Plants* (London: Methuen, 1958), p. 146.

[5] Ibid.

[6] Ibid., p. 147.

[7] Ibid.

[8] Ibid., pp. 148-149.

mies, tends to increase rather than decrease the likelihood of unfortunate oscillations in pest populations.[9]

Conservation biologists quickly put the hypothesis to work, even while the evidence for it remained largely intuitive, and arguments such as Commoner's became popular.[10] But when the MacArthur version of the hypothesis was submitted to empirical test, the data failed to support it unequivocally. In an influential 1975 paper, Daniel Goodman summarized the mounting evidence against the hypothesis by responding to Elton's reasons for accepting it:[11]

(1),(2) Elton's first two reasons support the hypothesis only if it can be shown that mathematical models of, and laboratory experiments concerning, more diverse and complex systems prove them to be more stable than simple systems. In fact, computer models with increased numbers of species showed greater instability.[12]

(3) The data suggesting vulnerability of islands to invasions by pest species may result from accidents of distribution or other special characteristics of islands.[13]

(4),(6) Crops and orchard trees planted in pure stands do not represent experiments that have been allowed to run their course. Subsequent developments might result in a stable, though less diverse, system if tillage were to stop. Even if true experiments on monocultural plantings were performed, the problem regarding controls for the comparison mentioned regarding (1) would apply.[14]

(5) Diverse tropical biota are so complex that we do not fully understand them and large fluctuations may go unnoticed. Further, Elton's largely ancedotal data here must be set against countervailing anecdotes of major instabilities in some tropi-

[9] Ibid., pp. 149-150.
[10] Elton, for example, cites the above evidence for the hypothesis in a chapter entitled "The Reasons for Conservation." Ibid., pp. 143-153.
[11] Daniel Goodman, "The Theory of Diversity-Stability Relationships in Ecology," *The Quarterly Review of Biology* 50 (1975): 237-266.
[12] Ibid., p. 238.
[13] Ibid., pp. 238-239.
[14] Ibid., p. 238.

cal forest systems. There is mounting evidence that the tropical forests are surprisingly susceptible to destruction as a result of anthropogenic disturbances.[15] Finally, recent data comparing tropical rivers and tropical lakes shows less resilience in the more diverse lake system.[16]

Although Goodman does not mention them, counterexamples also accumulated showing that some rather simple systems, such as salt marshes, are highly stable.

The mounting data questioning the diversity-stability hypothesis embarrassed conservationists, who then attempted to dissociate their activist policies from the now disreputable principle. For example, David Ehrenfeld says:

> [C]onservationists are provoked into exaggerating and distorting the alleged value of non-resources. The most vexing and embarrassing example for conservationists concerns the diversity-stability issue. . . . In our eagerness to demonstrate a present "value" for the magnificent, mature, and most diverse ecosystems of the world—the tropical rain and cloud forests, the coral reefs, the temperate zone deserts, etc.—we stressed the role they were playing in immediate stabilization of their environments. . . . This was a partial distortion that not only caused less attention to be paid to the real, long-term values of these ecosystems but also helped obscure, for a while, their extreme fragility in the face of human progress.[17]

Thus, conservationists recoiled from the hypothesis partly because it lost scientific respectability. But they also realized that the hypothesis implies that highly diverse systems are prime candidates for heavy exploitation by man: their diversity provides

[15] Ibid., p. 239.

[16] Thomas Zaret, "The Stability/Diversity Controversy: A Test of Hypotheses," *Ecology* 63 (1982): 721-731. Zaret's data contradict McNaughton's claim that diversity stabilizes ecosystem function. See S. J. McNaughton, "Diversity and Stability of Ecological Communities: A Comment on the Role of Empiricism in Ecology," *American Naturalist* 111 (1977): 515-525.

[17] David Ehrenfeld, "The Conservation of Non-Resources," *American Scientist* 64 (1976): 651-652.

them with stability in the face of human disturbances. Acting on this implication threatened the very goals conservationists espoused.

Despite these setbacks many environmentalists persist in believing that diversity and stability are related and that a proper understanding of the relationship would support preservationist policies.[18] This intuitive idea is nurtured by the commonplace observation that as humans develop tracts of land the ecosystems there simultaneously diminish in diversity and become less stable and less "ecologically healthy." I have already noted that the concept of diversity is multiply ambiguous.[19] That of stability is even more so, as will be seen. Thus several different formulations of this intuitive relationship are possible, and the choice to interpret the diversity-stability relationship as one between within-habitat diversity and the stability of such ecosystems through time rests mainly in MacArthur's precedent-setting speculations of 1955. It would be premature to abandon the diversity-stability relationship and its intuitive application to preservationist concerns simply because a single formulation of the relationship is unsupported by the data. It seems reasonable to consider alternative formulations, keeping in mind both the motivations of environmental protectionists and the data that undermined acceptance of the original formulation.

4.2 An Alternative Formulation of the Diversity-Stability Relationship

If the diversity-stability relationship is to be useful in supporting species preservation, it should correlate stability with *total* diversity. Diversity of systems has no direct bearing on preservationist concerns because, even in systems undisturbed by anthropogenic alterations, within-habitat diversity waxes and wanes as the sys-

---

[18] See Thomas Zaret, "Ecology and Epistemology," *Bulletin of the Ecological Society of America* 65 (1984). 4-7.
[19] See Section 2.2.

tem passes through various successional stages.[20] Is there a promising connection between total diversity and ecological stability?

In order to answer this question it will be necessary to examine closely the various meanings of stability. This concept can apply either to abiotic environments (where stability refers to the absence of significant fluctuations in temperature, rainfall, et cetera) or to biotic systems. Species preservationists would agree with MacArthur that the application to biotic systems is the relevant one. But even here alternative conceptions and measures abound. The various measures of stability can be organized into two useful categories; I will call them "static" and "dynamic" conceptions of stability.

Measures of either constancy (the ability of a system to resist change in response to disturbance) or resilience (the ability of a system to return to a "normal," that is, predisturbance, state subsequent to a disturbance) are static conceptions of stability. Constancy is obviously a static conception. But resilience is static, too. Resilience concepts describe the degree, manner, and pace in which the normal structure and function of a system are restored subsequent to disturbances.[21] Viewed over time, the effect of high degrees of resilience is to continue the system in an unchanged, static state.

Few ecosystems could be described as constant. Even systems that have powerful mechanisms for reacting to environmental fluctuations usually do so through internal changes that return the system as quickly as possible to a stable state. But these mechanisms involve important alterations and are better described as exemplifying resilience than constancy. Measures of resilience,

[20] See Section 3.2.
[21] Gary W. Harrison, "Stability under Environmental Stress: Resistance, Resilience, Persistence, and Variability," *American Naturalist* 113 (1979): 660; Walter E. Westman, "Measuring the Inertia and Resilience of Ecosystems," *Bioscience* 28 (1978): 705; C. S. Holling, "Resilience and Stability of Ecological Systems," *Annual Review of Ecology and Systematics* 4 (1973): 1-23; Gordon H. Orians, "Diversity, Stability, and Maturity in Natural Ecosystems," in *Unified Concepts in Ecology*, ed. W. H. Van Dobben and R. H. Lowe-McConnell (The Hague: Dr. W. Junk, B. V. Publishers, 1975), pp. 139-149; and Robert M. May, *Stability and Complexity in Model Ecosystems* (Princeton, N.J.: Princeton University Press, 1973).

unlike those of constancy, recognized the way ecosystems maintain long-term normalcy by reacting to ("damping out") disturbances with short-term alterations. It is a further advantage of these measures that they correspond to Lyapunov stability as studied by mathematicians. Consequently, measures of resilience are well understood, easy to test for, and mathematically tractable. It is this conception of stability that was used by MacArthur in his original formulation, and it has become the standard point of reference for discussions of the diversity-stability hypothesis.

In spite of these advantages ecologists are coming to recognize that this family of concepts has limited applicability to actual ecosystems.[22] Lyapunov stability measures a system's ability to return to a single stable point in response to a disturbance. In fact, most ecological systems appear to exhibit multiple points of stability. If a disturbance remains below a certain threshold (which varies with respect to different characteristics and types of disturbances as well as across systems), the system will return to its predisturbance level. This is called neighborhood stability. But if the magnitude of the disturbance exceeds the threshold for that disturbance, the system will find a new point: it has been driven into a new "basin of attraction." Each of these points exhibits neighborhood stability (that is, the system will return to this condition provided a disturbance is not too severe), but the system is not globally stable.[23] Unfortunately, empirical as well as mathematical difficulties are involved in determining whether systems that exhibit neighborhood stability are also globally stable.[24] Resilience, as measured by Lyapunov stability, then, is appropriate to systems only insofar as they are, and are known to be, globally

[22] Richard C. Lewontin, "The Meaning of Stability," in *Diversity and Stability in Ecological Systems*, ed. G. M. Woodwell and H. H. Smith, in *Brookhaven National Laboratory Publication No. 22* (Springfield, Va.: Clearinghouse for Federal Scientific and Technical Information, 1969); Holling, "Resilience and Stability"; G. Innis, "Stability, Sensitivity, Resilience, Persistence. What Is of Interest?" pp. 131-139, and L. Wu, "On the Stability of Ecosystems," pp. 155-165, both in *Ecosystem Analysis and Prediction*, ed. S. Levin (Philadelphia: Society for Industrial and Applied Mathematics, 1974); and Orians, "Diversity, Stability, and Maturity."
[23] Lewontin, "The Meaning of Stability," p. 15.
[24] Ibid., pp. 16, 22.

stable. But ecological systems are not globally stable. Therefore, however convenient it is to employ Lyapunov stability in studying such systems, that measure is inappropriate unless one has independent evidence determining the magnitude of disturbance thresholds with respect to the various characteristics involved. Since much discussion of stability must be directed at determining such thresholds, the concept of resilience will often be inapplicable.

Further, systems do not change only in response to disturbances. They also undergo natural, successional change through time. While theorists differ in their account of the mechanisms whereby succession takes place, it is widely agreed that it is normal and natural for systems to change as a function of both ecological and evolutionary time.[25] Insofar as constancy and resilience are both static conceptions of stability, they seem only to apply to systems frozen in time. Static conceptions of stability fail to provide a framework within which to examine adaptations of species to other species in ecosystem devlopment or genetic adaptations of species through evolution.[26]

The traditional formulation of the diversity-stability hypothesis is, then, unfortunate in several respects. The choice to correlate within-habitat diversity with stability limits the applicability of the hypothesis to concerns of species preservationists, who are not concerned with the normal disappearance of species from evolving systems. The choice to use static measures is likewise unfortunate because ecosystems are dynamic in nature.

Systems can also be described as dynamically stable, however. Although he does not use the term "stability," Ramon Margalef gives an apt description of this concept:

The conclusion is that in any estimate of maturity, not only diversity, but also predictability of change with time has to

[25] Joseph H. Connell and Ralph O. Slatyer, "Mechanisms of Succession in Natural Communities and Their Roles in Community Stability and Organization," *American Naturalist* 111, no. 982 (1977): 1,119-1,143.
[26] Geerat Vermeij, personal communication. See Orie Loucks, "Evolution of Diversity, Efficiency, and Community Stability," *American Zoologist* 10 (1970): 17-25; and Innis, "Stability, Sensitivity, Resilience," pp. 131-139.

be considered. Ordinarily both characters are correlated. Less mature ecosystems not only have a lower diversity, but in their transition between successive states include a higher amount of uncertainty. And more diverse ecosystems have, in general, more predictable future states. In other words, in more mature ecosystems the future situation is more dependent on the present than it is on inputs coming from the outside. Homeostasis is higher. On the other hand, future states in less mature ecosystems are heavily influenced by external inputs, by changes in the physical environment.[27]

While Margalef's remarks on diversity of systems would now be disputed by many population biologists, our present concern is with the concept of stability. A system is dynamically stable, according to Margalef, if its future states are predictable largely from its present state, with little reference to outside influences. Dynamic stability refers to the extent that the system is "autogenic," that is, to the extent that its internal states result from forces internal to the system rather than from outside forces. This conception of stability carries no implication that a stable system must remain unchanged through time. Some species may become more abundant as others decline and even disappear without any implication of instability.

A dynamic conception of stability is not without problems, of course. First, it might be thought that dynamic stability is a contradiction in terms, as the former term suggests change while the latter suggests constancy. But this worry is based on a confusion. A changing system can be stable, provided different features of the system are referred to by each label. There is no contradiction in saying that an automobile hurtling down the highway is stable. Its forward momentum represents change, while its resistance to sideways or twisting motions represents stability.

Second, the adoption of a dynamic conception of stability requires abandonment of Lyapunov stability as understood by

[27] Ramon Margalef, "On Certain Unifying Principles in Ecology," *American Naturalist* 97 (1963). Also see Eugene P. Odum, "The Strategy of Ecosystem Development," *Science* 164 (1969): 262-270.

mathematicians because dynamic stability cannot be related to a single stable point—it is relativized to degree and type of disturbance. The use of a dynamic conception of stability thus causes some inconveniences; but this may be the price of recognizing the inapplicability of static conceptions of stability to actual ecosystems.

Third, it is not obvious which characteristics to choose as the indicator of dynamic stability. Static stability is normally related to persistence and relative abundance of species. But dynamic stability cannot be tied to such factors because species in dynamic systems will naturally vary in their abundance and even in whether they persist across time, according to the stage of ecosystem development. Therefore, species persistence and abundance cannot be an indicator of dynamic stability.

This may appear very damaging. But it must be remembered that, whatever characteristic is chosen to be measured—as when ecologists choose to measure static stability in terms of species abundances—it is *chosen* as an *indicator*. When ecologists speak of stability, they mean ecosystem stability, and thus refer to a wide range of ecosystem characteristics. For the purposes of convenient scientific observation they choose some single characteristic that is easily measured and at the same time is a fairly reliable indicator of the whole range of characteristics that are included in the intuitive idea of stability.[28] The return of some specific species abundance to a predisturbance state was thought to be a useful indicator of stability across time. But I have just argued in essence that such an indicator falsifies the situation by freezing in time systems that are in fact dynamic.

What characteristic should, then, be chosen as the best indicator of dynamic stability? This question is not easily answered, as a choice can be made only after careful observation and compar-

[28] Indeed, MacArthur's original formulation of the diversity-stability hypothesis arose in just this way. He thought that diversity would be more easily measured than stability, and the point of his hypothesis was to suggest that a diversity measure could be used as an indicator of stability. See MacArthur, "Fluctuations of Animal Populations," and Goodman, "Diversity-Stability Relationships in Ecology," p. 239.

ison of a wide range of ecosystems. It would be beyond the scope of this essentially nonempirical study to propose some particular indicator. The argument here is better seen as placing two constraints upon any concept chosen as an indicator of ecosystem stability. First, such an indicator must not preclude the possibility that a system can be described as stable while changing through time. Second, it must measure the intuitive idea that any changes occurring result from factors internal to the system rather than from external causes because dynamic stability measures the internal predictability of systems.

To summarize, I have suggested that the diversity-stability hypothesis is not a single, well-defined principle but rather an informal idea that some meaningful relationship exists between diversity and stability, two highly ambiguous concepts. It was a historical accident that MacArthur chose to formulate the relationship as correlating within-habitat diversity with resilience of ecosystems. Species preservationists quickly adopted that hypothesis as supporting their cause, only to discover that as the evidence came in, the hypothesis was highly questionable. Worse, it seemed to suggest management directives contrary to their own goals. They retreated in embarrassment, abandoning attempts to appeal to that relationship to support their policies.

But this may have been an overreaction. Careful analysis shows that within-habitat diversity is not a prime candidate to illuminate preservationists' concerns, for it focuses on static levels of species' populations within individual ecosystems, while preservationists are normally concerned with total diversity and dynamic stability. With these observations in mind, I am suggesting that species preservationists should propose an alternative formulation of the intuitive notion of a link between diversity and stability. In particular, total diversity, the total number of species existent within a geographical area, may well be correlated with dynamic, or autogenic, stability.

Intuitively this formulation of the diversity-stability hypothesis emphasizes the importance of a large number of species existing in varied ecosystems across an area. These species are potential invaders and colonizers of open spaces so that, following a disturbance such as a forest fire, the numerous species representing the high degree of total diversity in the area will begin an intense competition for niche space in the open area. A minimal number of species is necessary for this process to proceed naturally, and the larger the number, the more intense will be the ensuing competition. The proposed version of the diversity-stability hypothesis claims that, when ecosystems develop under such intense competition, they will attain, both more quickly and to a greater degree, the characteristics associated with dynamic stability. Systems in which competition for niche space is intense will develop a high degree of complexity, interrelatedness, and niche specialization. These characteristics lead to greater degrees of autogenic determination: alterations of the system become increasingly determined by features internal to it.

This relationship is in one sense paradoxical. It is being suggested that as external forces impinge on a particular ecosystem, the system becomes less determined by those forces. But this is only an apparent paradox. The more numerous and varied are the invading species initially, the more quickly and more fully will defenses against such invasions develop. Systems that evolve in areas of high total diversity will contain more specialization, more symbiotic relationships, more interdependencies on other species in the system. The development of these complex relationships will decrease dependence upon outside forces and the system will resist changes in response to causal factors impinging from the outside. Having endured many invasions and either incorporated or repulsed the invading species, the system is more prepared than before to resist future invasions and disturbances.

4.3 Dymamic Stability and Autogenic Systems

Versions of the diversity-stability hypothesis patterned on MacArthur's speculative precedent can now be seen to describe an important relationship too simply. MacArthur's hypothesis connects ecosystem diversity directly with ecosystem stability. I am suggesting, instead, that ecosystem diversity increases and decreases quite naturally as various stages of ecosystem development unfold. The important variable affecting ecosystem stability is "maturity" (time elapsed since the last major disturbance). During this time, species invade, colonize, and develop more specialized competitive as well as symbiotic relationships.[29] It is these relationships that create the sort of dynamic stability that mature ecosystems exhibit.

But that stability is not directly correlated with ecosystem diversity. While ecosystems initially tend toward greater diversity (which accounts for the intuitive plausibility of the MacArthur version), there is no guarantee that intense competition for niche space must result in more and more species within a system. If a small number of highly specialized species work out effective mechanisms for exploiting an abiotic environment, they may exclude others, creating a simple but productive system. The important variable is not diversity *within* the systems in question but the backdrop against which these species compete for niche space in the system—the total diversity that provides the stock of potential invading and colonizing species. Consequently ecosystem stability should be correlated with areawide total diversity: we can expect that ecosystems emerging in highly diverse areas will exhibit the greater specialization of function and interconnectedness of relationships characteristic of highly mature systems. These in turn are responsible for the autogenic stability observed in mature systems.

This more complex relationship between diversity and stability is consistent with the observation that some mature and stable systems, such as salt marshes and redwood forests, are not diverse. These are systems where intense competition has produced a set of species so specialized to their particular biotic environment and to the surrounding species with whom they have evolved symbiotic and competitive relationships that they have achieved dominance over large portions of their range. The success of these associations is a function of past competition among a large number of species against a backdrop provided by the abiotic conditions in which the competition took place. These systems are mature and autogenic, hence, stable. The stability of

[29] These processes are described in detail in Section 3.3.

such simple systems is no less dependent upon the initial total diversity from which the competitors are drawn than is that of highly diverse systems.

A common objection to preservationist arguments can thus be dispatched. It is often thought that species preservationists argue for protection of wilderness and other relatively undisturbed areas because more mature systems are more diverse. To this critics respond that maturity does not always result in diversity, as we have just seen. But this objection mistakenly assumes that the preservationists' case is built on the importance of within-habitat diversity. When total diversity is seen as the focus of preservationist arguments, the critics' attack misses its mark. Total diversity is enhanced in an area when there are many types of systems at many different stages of development. Mature systems make a special contribution to total diversity because they contain highly specialized and highly interconnected species in complex relationships that have not yet evolved in less mature systems. Since many areas of the earth's surface have now been altered for intense human use, there are fewer places where these most specialized species can exist. To eliminate the least disturbed areas is to eliminate the habitats of species that exist nowhere else. One need not claim that mature ecosystems are the most diverse to claim that they, by containing species that exist there and nowhere else, contribute to total diversity.

Our version of the diversity-stability hypothesis can accommodate some often-cited and much misunderstood data from analytical mathematics. Robert May and others found that adding elements to randomly organized simulated systems decreases rather than increases their stability (resilience).[30] These results seem at first to undermine the diversity-stability hypothesis, but more recent publications have placed these mathematical findings in a different perspective. Several authors, including May himself, have recognized that naturally occurring ecosystems form a very small subset of mathematically possible systems and that all members of this subset share the characteristic of not being or-

[30] May, *Stability and Complexity in Model Ecosystems.*

ganized randomly.[31] If biologically complex systems are more stable, this stability would result from their having evolved highly nonrandom interrelationships, as just explained. Both resilience and, especially, dynamic stability seem to be increased by greater complexity, especially if this greater complexity involves increased interconnectedness of the species involved.[32] Thus, mathematical data are consistent with a version of the diversity-stability hypothesis connecting total diversity to dynamic stability through the intermediary of highly developed interconnectedness relationships.

The proposed relationship between total diversity and ecosystem stability also does not run afoul of observations of the fragility of tropical forests, temperate deserts, and other diverse systems. Indeed, the proposed hypothesis helps provide a reasonable explanation of that fragility. Tropical forest ecosystems, when mature, are highly interconnected and highly interrelated. The intense competition resulting from the overwhelming total diversity in the tropics has created high degrees of specialization and highly organized, efficient systems there. In the process they have evolved mechanisms for damping out disturbances of average magnitude. Had they not developed these mechanisms, they would be unstable, continuing to evolve as species that take advantage of such usual disturbances fight for a role in the system. But the same processes that increase the effectiveness of the system in dealing with standard and expected disturbances make them more susceptible to new, more intense, and more pervasive ones. Greater local stability is bought at the price of lesser global stability because the finely tuned interconnections and balanced competitions that produce stability in the face of normal disturbances also create dependencies of species upon species.[33]

[31] See, for example, Robert M. May, *Theoretical Ecology* (Philadelphia: W. B. Saunders Co., 1976), pp. 159-162; and Lawrence R. Lawlor, "Structure and Stability in Natural and Randomly Constructed Competitive Communities," *American Naturalist* 116 (1980): 394-408.

[32] D. L. De Angelis, "Stability and Connectance in Food Web Models," *Ecology* 56 (1975): 238-243.

[33] D. Futuyma, "Community Structure and Stability in Constant Environments," *American Naturalist* 107 (1973): 443-446.

If a major or new type of disturbance such as those introduced by humans suddenly impinges on the system, these relationships are thrown out of kilter. These alterations cause abrupt extirpations whose consequences cascade through the system, causing further disruptions. The result can be a total breakdown of the system.[34] Once local and global stability are distinguished, the proposed hypothesis predicts this result. If total diversity in an area contributes to greater interconnectedness among species within systems and if greater interconnectedness leads to greater ability to damp out normal disturbances and lesser ability to respond to major new disturbances, this explains the fragility of highly diverse and interconnected systems.

Finally, the proposed hypothesis does not support the dangerous management practices that seemed to follow from the principle that ecosystem diversity produces ecosystem stability.[35] It refutes the suggestion that diverse and interconnected systems should be able to absorb higher degrees of human exploitation because their diversity makes them more stable. Mature and highly integrated systems would be particularly threatened by human exploitation because their high degree of local stability belies their global instability. When humans introduce alterations that have no analogues in the normal environmental fluctuations affecting systems, they should expect significant, pervasive, and often unfortunate results.

I have so far introduced and explained an alternative version of the intuitive connection between diversity and stability. I have not offered new data to support this version. I have not even suggested a characteristic of ecosystems that could be measured as an indicator of dynamic, autogenic stability. I have, however, attempted to show that the new version avoids some of the problems associated with MacArthur's original one, and I have argued that it is consistent with and even explains some important data about the effects of alterations on ecosystems. Some scientific data also lend plausibility to the hypothesis. I am referring to the

[34] See Section 3.3.C.
[35] Ehrenfeld, "The Conservation of Non-Resources," pp. 651-652. See quotation above, at note 17.

studies cited in Section 3.3.A, which imply that systems will develop more quickly in areas with greater total diversity. There it was claimed that secondary succession (characterized by the presence of seeds and propagules) progresses more quickly than primary succession (where seeds and propagules are not present). Also, Cody's studies show that accessibility of a habitat to a large pool of potential colonists increases between-habitat diversity, and studies in island biogeography indicate that colonist accessibility affects between-habitat diversity. Greater between-habitat diversity can be seen as an indicator that biotic factors determine ecological succession because greater variety in vegetational structure creates microhabitats and more opportunities for specialists. In a hardwood forest there will be an overstory, an understory, and the forest floor, and all provide potential niches. The competition for these niches, waged in the context of normal abiotic fluctuations, can be expected to create an autogenically determined system. Therefore, while no new studies have been done to test this proposed version of the diversity-stability hypothesis, it does square with existing data.

The new version of the hypothesis requires, of course, clear and operational statement, scientific analysis, and testing. These tasks must be left to the scientific community, as they lie beyond the scope and intent of this book. For the remainder of this chapter I will provisionally accept the proposed version and examine some of the consequences that follow from it, if it is true.

4.4 Autogenic Systems and Human Benefits

In the last section it was shown that a reasonable, though largely untested, hypothesis suggests that a high degree of areawide total diversity is a necessary condition for the emergence of highly complex and mature ecosystems. The hypothesis also implies that greater degrees of total diversity provide conditions favoring greater dynamic stability. In this section I will show how the existence of such autogenic systems is important to human well-being.

Argument 4

Systems that develop in natural successional sequences offer a variety of benefits to humans. But these benefits are not easily recognized or measured, in contrast to the easily quantifiable benefits that accrue from less natural land uses. In order to explain this point I will contrast the advantages of a monocultural agriculture with those of a more natural, evolved, and highly organized system.[36] Monocultures can be extremely productive. Species important as agricultural crops are opportunistic species that turn their energy into reproduction, wasting little on building structure in the form of woody stems, et cetera. So, in one sense, modern agriculture maintains its productivity by holding back successional development and by exploiting the reproductive energy spent by opportunistic species that produce fruit or grain. But such high productivity is costly in the sense that harvesting removes energy from the system. Even more crucial is that the food web is not closed in the sense that little primary production finds its way back into the store of nutrients available; thus, highly cultivated monocultures are usually kept in their highly productive state only by importing great amounts of energy from outside the system. First, they require energy (work) to maintain them at the early successional stage. Today this energy comes increasingly from fossil fuels. Second, they require massive inputs of fertilizer in order to maintain production as harvesting removes energy and the open nutrient cycle leads eventually to nutrient impoverishment.

These points are well known. The important thing to emphasize is that artificial production and maintenance of monocultures require escalating inputs of energy from outside the system. The family farm, raising several species of crops and animals and making fuller use of energy, could mimic nature by recycling nutrients in the form of chaff and manure. While productive levels on family farms are lower, they tend to be more efficient in the on-

[36] The following discussion is similar to that of Paul Colinvaux in *Introduction to Ecology* (New York: John Wiley and Sons, Inc., 1973), p. 240.

site production and use of energy. It follows that they are more efficient *in the long run*, taking into account the inevitable exhaustion of nonrenewable resources used for fuel and chemical fertilizers. Systems that are less disturbed than monocultures will contain collections of species that have coevolved in the local environment. These collections of species are more likely to be better at exploiting the resources of that environment than are the generalist species promoted in monocultures.[37]

In a world increasingly dependent upon monocultures, there will be not only a decrease in the area covered by complex, naturally evolved, and efficient systems closed in their nutrient cycles but also a decrease in the possibility of such systems emerging in the future. If present trends continue, monocultures will cover much of the land area of the globe, and the local and worldwide extinctions caused by such practices will undermine the total diversity of local areas and of the world as a whole. The resulting impoverished world will lack the building blocks for the reconstruction of such systems. The use of humanly enforced monocultures produces a spiraling trend: as more and more space is so occupied, there will be less and less possibility of reversing the trend by reinstituting unmanaged areas and allowing them to evolve into complex, efficient systems. This spiral, of course, leads to yet another spiral: more and more energy must be imported into productive farm lands, causing ever-growing demands for energy and artificially produced nutrients.

I am suggesting, then, that maintaining total diversity has great value because it provides the only means of retreat from the present tendency toward simplification. A "patchy" area, containing a variety of ecosystems at differing stages of development, will have greater total diversity. When land in monocultures loses productivity it can lie fallow and, if there are enough specialist species available to develop a productive and closed ecosystem, nutrients will be restored to the soil and it will become productive once again. I refer to these advantages as the regenerative value of

[37] Lawrence B. Slobodkin and Howard L. Sanders, "On the Contribution of Environmental Predictability to Species Diversity," in *Diversity and Stability in Ecological Systems*, p. 85.

ecosystems and indirectly of species. If only opportunistic species such as weeds, rats, and cockroaches remain, development will be slow, as in abandoned areas in the center of large cities.

Total diversity of areas, then, provides extremely valuable insurance against downward spirals in productivity. An ideally productive environment would have patches of monoculture interspersed with natural systems. As uses are rotated, species will move from one area to another. They can, therefore, contribute to the rejuvenation of areas where intense use leads to deterioration. Such a situation has some hope of sustaining productivity without overtaxing nonrenewable resources such as chemical nutrients and fossil fuels. The economic benefits of turning mature systems into highly productive monocultures are both immediate and easily measured. But these benefits are, in a sense, misleading. They are owed to massive infusions of fossil-fuel energy and nutrients created eons ago by the very systems the monocultures are replacing. The benefits of a highly developed, efficient system with a closed food chain are "free" in the very sense that the highly managed systems are "expensive." Using up nonrenewables at unnecessarily rapid rates is, essentially, spending capital. Those resources will increase in value in the future, and using them now to achieve short-term maximal production undermines future productivity for present economic gain.

Further, dynamic stability itself affords significant benefits that are not easily recognized or measured. Maturing systems develop complex interrelationships among species in response to predictable and rhythmic changes in the environment. But in the same process, the environment itself is controlled. It is well known, for example, that a heavy vegetative cover encourages rainfall. On a less global scale, forest canopies equilibrate temperatures on the ground below, presenting attractively cool homesites for humans and other creatures. In general, mature systems are more dependent on internal, biotic factors than on abiotic factors. Consequently, such systems mitigate environmental harshness (sometimes in areas larger than themselves). Obvious examples where failures to maintain a complex abiotic community have increased environmental harshness are the dust bowl phenomenon, where

92

interacting factors including loss of windbreaks and vegetative cover led to devastating dust storms, and expansions of the desert, where over-cultivation and overgrazing led to major biotic and abiotic changes.

It is also important to note, highlighting points made in Section 4.2, that highly developed, autogenic systems provide more ecosystem services than do simple ones.[38]

Finally, highly organized systems promote new symbiotic relationships between species and the development of new highly specialized and efficient species. While speciation in general may take place more quickly in immature systems (because the life cycles of the organisms in such systems are generally short, among other reasons), this speciation is unlikely to result in species highly specialized to particular environmental niches. Species evolving in less complex systems will be generalists, not specialists. The long-term development of more specialized species is especially valuable to humans. This point is sufficiently important to warrant a separate section.

4.5 Selective Extinction and Human Benefits

In demonstrating the prima facie value of each and every individual species, I have so far emphasized characteristics common to all species. I have tried to show the prima facie value of any species, regardless of its particular characteristics. I want now to concentrate on characteristics shared by extremely broad groupings of species. Scientists agree that species are not equally susceptible to extinction from human-initiated causes.[39] It turns out that

[38] Odum, "The Strategy of Ecosystem Development," p. 269.

[39] E. O. Willis, "Populations and Local Extinctions of Birds on Barro Colorado Island, Panama," *Ecological Monographs* 44 (1974): 153-169; John Terborgh and Blair Winter, "Some Causes of Extinction," in *Conservation Biology: An Evolutionary Ecological Perspective*, ed. M. E. Soule and B. A. Wilcox (Sunderland, Mass.: Sinauer Assoc., 1980), pp. 119-133; Robert E. Jenkins, "Endangerable Species," *Ecology Forum* 25 (Fall 1977): 20-21; Lawrence Slobodkin, "On the Susceptibility of Different Species to Extinction: Elementary Instructions for Owners of a World," in *The Preservation of Species*, ed. Bryan G. Norton (Princeton, N.J.: Princeton University Press, 1986).

those species that are most susceptible to extinction are the very species that are characteristic of the highly complex, mature, and predictable systems described in earlier sections of this chapter. Further, the general characteristics that make species most susceptible to human-caused extinction also make them most likely to be useful to humans. Therefore the prima facie value of each species can actually be multiplied in considerations of endangered species policy.

Argument 5

The principal human activities that endanger species are habitat destruction and fragmentation, hunting, air pollution, and the introduction of foreign species. Geerat Vermeij has concluded that habitat destruction, fragmentation, and hunting are the chief culprits in the immediate wave of extinctions and endangerments.[40] Which species are most likely to be threatened by these human activities? There seems broad agreement that the following factors increase likelihood of endangerment or extinction: (1) rarity (either sparse distribution over a wide range or confinement to narrow range);[41] (2) large individual size;[42] (3) high trophic level;[43] (4) biotically controlled evolution;[44] (5) low dispersibility, few offspring, and long individual life spans (K-selected species);[45] (6) specialization of habitat;[46] (7) involvement in mutual-

[40] Geerat J. Vermeij, "The Biology of Human-Caused Extinction," in *The Preservation of Species*.
[41] Terborgh and Winter, "Some Causes of Extinction," p. 128; Jenkins, "Endangerable Species," p. 21; Charles W. Fowler and James A. McMahon, "Selective Extinction and Speciation: Their Influence on the Structure and Functioning of Communities and Ecosystems," *American Naturalist* 119 (1982): 482.
[42] Jenkins, "Endangerable Species," p. 21.
[43] Ibid., p. 21; Fowler and McMahon, "Selective Extinction," p. 483.
[44] Geerat J. Vermeij, *Biogeography and Adaptation* (Cambridge, Mass.: Harvard University Press, 1978), p. 183; Vermeij, "The Biology of Human-Caused Extinction."
[45] Robert H. MacArthur and Edward O. Wilson, *The Theory of Island Biogeography* (Princeton, N.J.: Princeton University Press, 1967), p. 151; Vermeij, *Biogeography*, pp. 186-187.
[46] Terborgh and Winter, "Some Causes of Extinction," p. 21.

isms and coevolutionary arrangements;[47] (8) existence in ecosystems of high diversity.[48]

These characteristics are by no means independent of one another, as some of them tend to co-occur. K-selected species invest relatively small amounts of energy in reproduction; they have low dispersibility because they have fewer offspring. They also tend to have large body size and so tend to live longer lives, since large size is a comparative advantage in many competitive relationships.[49] These characteristics are, in turn, correlated with another: many K-selected species are "biotically competent." That is, they evolve largely in response to adaptational pressures from other organisms rather than in response to abiotic factors such as climate, soil conditions, et cetera. Thus, such species are also more involved in mutualisms and coevolutionary arrangements. They are also likely to exist in specialized habitats characteristic of the highly diverse, interconnected, mature ecosystems discussed in the last section. In these ecosystems species at the upper end of relatively long food chains are most susceptible to extinction.[50] The inefficiency of energy transfer from trophic level to trophic level ensures that these upper-level species will be comparatively rare, and their dependency on lower levels also increases susceptibility.[51]

There are several reasons why species with these characteristics are also more likely to be important to humans. First, the loss of species from highly interrelated systems is more likely to cause a cascade of further extinctions. While threatened species are likely to be rare, they may be important over a narrow range because

[47] Fowler and McMahon, "Selective Extinction," p. 483.

[48] Kenneth J. Hsu et al., "Mass Mortality and Its Environmental and Evolutionary Consequences," *Science* 216 (1982): 255; Orians, "Diversity, Stability, and Maturity."

[49] J. P. Grime, "Evidence for the Existence of Three Primary Strategies in Plants and Its Relevance to Ecological and Evolutionary Theory," *American Naturalist* 111 (1977): 1,189; Fowler and McMahon, "Selective Extinction," p. 485.

[50] Terborgh and Winter, "Some Causes of Extinction," pp. 119-133.

[51] Lawrence Slobodkin, "Ecological Energy Relationships at the Population Level," *American Naturalist* 95 (1960): 213-236.

they are likely to be highly integrated with other species. Thus, the very fact that a species is threatened implies, by virtue of the correlations just developed, that its extinction is more likely than that of a randomly chosen species to contribute to ecosystem deterioration and that its preservation is of special importance in protecting humans from the effects of catastrophic ecosystem breakdown. Humans, being high on the food chain, are dependent on a number of lower links. Disturbances that sever those links are likely to affect human interests in unforeseen, and likely unfortunate, ways.[52]

Second, the specialized, biotically competent species just described as more susceptible to extinction are also more useful and "interesting" for human purposes.[53] If each species is viewed as a solution to some set of environmental problems, the biotically competent species are those that have solved the most interesting and complex ones because they have evolved in response to a complex, diverse, and highly interconnected system. For example, most of the substances that have yielded medicines useful to humans evolved as chemical defense mechanisms of plants against intense predatory pressures. The humpback whale has solved the problem, shared by humans, of long-range underwater communication. Current research is designed to understand and perhaps apply those solutions to human problems.[54]

Generalist species that survive by high dispersibility and live short lives are much less in need of such defenses and artifices. These are the pestiferous and weedy species that make human lives less pleasant. Insofar as they have solved environmental problems, they have resorted to a few simple strategies, and they offer far less fertile ground for research for producing useful goods and services.

Third, like the susceptible species under discussion, humans are K-selected, that is, have comparatively large body size, are high in trophic level, exist in highly complex, even social, systems, et cet-

[52] Fowler and McMahon, "Selective Extinction," p. 494.
[53] Vermeij, "The Biology of Human-Caused Extinction."
[54] Terry Leitzell, "Species Protection and Management Decisions in an Uncertain World," in *The Preservation of Species*.

96

era. Thus, it is not surprising that those susceptible species are useful for understanding human life and human society. The highly evolved species that share characteristics (1) through (8) are more likely to be studied and tested in experiments concerning human health. Likewise, the complex social relationships represented within and between species of this sort have far closer analogies to human society than those of species not biotically evolved.

When a biotically competent, specialist species is extirpated from a system, its niche is usually not filled by another similar species. Such species are too specialized to allow interchangeability. Instead, they are likely to be replaced by generalist, weedy species that colonize quickly and achieve ready dominance. Over long periods of time with no further disturbances, more specialization may occur. But a dangerous downward spiral in species diversity may have begun. As a study of mass extinctions in the past shows, recovery from serious breakdowns of this sort can take millennia.[55] In the meantime, if humans can survive at all, they will be forced to share the planet with comparatively dull, generalist, weedy, and pestiferous species.

Thus, each species has a considerable prima facie value. But humans must also realize, when considering the possible loss of a species, that the very characteristics that increase vulnerability to extinction also increase the likelihood that a species is useful and attractive to humans. While species have considerable prima facie value regardless of their special characteristics, threatened species have an even greater value because of the increased probability that they will have still further special characteristics of interest to humans.

[55] Hsu et al., "Mass Mortality," p. 255; Robert K. Peet, "Ecosystem Convergence," *American Naturalist* 112 (1978): 442.

FIVE

AMENITY VALUES

5.1 Necessities and Amenities

Demand values represent aggregations of felt preferences, temporarily frozen in time for the purpose of computation. Many of them are not, of course, actually fixed either genetically or physiologically. There is therefore an intuitive but by no means sharp distinction between preferences that are based on essential human "needs" and those that are changeable and culturally or psychologically determined ("wants"). Attaining a minimum caloric intake, protection from life-threatening predators or parasites, and maintenance of a minimum body temperature provide relatively clear examples of essential needs, while the desire for symphonic music, for meat at every meal, and for a fur coat are examples of nonessential felt preferences.

This distinction, however clear in cases such as these, marks no sharp dividing line. Even the mentioned examples require qualifications under special circumstances: in Eskimo cultures without trade connections to less arctic climates, for example, fur coats may, as the only practical means to maintain a minimum body temperature, become necessities. Nor can minimum provisions for physical survival be taken as the only "essential" human demands. People do not flourish and maintain robust strength and good health on minimal caloric intakes. Some diversity of foodstuffs is essential to good nutrition. When it is further recognized that humans are highly social animals by nature, a case can be made that the types of food eaten and even preparatory practices are "culturally necessary." The difficulty in drawing such lines as these is illustrated by the ongoing controversy between Jews who value kosher customs of slaughtering domestic animals and ani-

98

mal rights advocates who argue for more humane techniques of killing animals for human consumption.

In spite of the hazards involved in distinguishing "necessities" from nonessential desires, or "amenities," it is customary to divide demands according to some such rough-and-ready subcategorization. According to this division, a range of reasonable means to fulfill basic needs are counted as necessities, while extravagant means to fulfill basic needs (gluttonous feasts and gold-embroidered clothing) are combined with standard means to fulfill less essential human demands to form the category of amenities. The discussion of Part A has thus far concentrated on examples where wild species fulfill essential human demands, such as those for food, medicine, and shelter. I have argued that accurate monetary assessment of these essential demand values remains elusive. In turning now to a discussion of demands for amenity values derived from wild species, similar problems confound accurate monetary assessments but, in addition, special problems attendant upon the valuation of amenity values arise. First, some types of amenities are seldom, if ever, purchased in markets, and the role of contingent valuation is correspondingly greater for amenities than for commodities and services. These problems will be discussed in Section 5.2. Second, several special features of amenity values tend to frustrate accurate valuation. These special features will be examined in sections 5.3 through 5.5.

5.2 Contingent Valuation of Amenities

Lists of amenities derived from wild species and natural ecosystems include: recreation such as bird watching, hiking, and canoeing;[1] the joy of scientific learning and understanding that

---

[1] Perhaps this list should also include hunting and fishing, although these uses are obviously problematic when the focus is on endangered species. Still, a diverse biota and less developed areas are essential to support game species, and hunters and fishermen are among the strongest supporters of wildlife protection efforts. See Judd Hammack and Gardner Mallard Brown, *Waterfowl and Wetlands: Toward Bioeconomic Analysis* (Baltimore: Resources for the Future, 1974), p. 3. Hammack and Brown state that only consumptive uses depend upon comparatively large numbers of waterfowl (they assert that numbers adequate for viewing

often accompanies nature hikes and naturalist expeditions; aesthetic values derived from viewing natural ecosystems and wild species; appreciation of the religious, symbolic, and historical significance of wild species and their contribution to cultural experiences; the experience of solitude possible in wilderness and the therapeutic and character-building possibilities attendant upon interactions with wild species in natural places.[2] Values derived from these experiences normally depend directly upon relatively unspoiled natural habitats. They therefore depend indirectly on the wild species of which these habitats are composed. In some cases, highly valued experiences are directly attributable to an endangered species, as when a bird watcher gains special pleasure from viewing a rare whooping crane.

Markets exist for some amenities derived from natural species. For example, safaris and canoe trips can be purchased for fees. When one stands at a scenic overlook beside a highway, there is often an opportunity to purchase twenty-five cents worth of view through a telescope. But in most cases amenities derived from wild species and natural ecosystems are free or require only nominal fees because of government subsidization of recreational facilities. Some indication of the demand for such experiences can be obtained by noting the amount hikers pay for gas to drive to the hiking trail, for lodging, gear, et cetera when they take advantage of recreational and other opportunities. But most visitors to national parks, forests, and wilderness areas insist their experiences are more valuable than the small fees they pay.[3] Consequently, these amenities are "contingently" valued; a price is assigned according to hypothetical market techniques. Questionnaires are designed to determine how much users would be will-

and photography would exist without protective efforts), and thus only hunting will justify values on additional waterfowl.

[2] This list includes the relevant subset of a more inclusive list of "wildland values" as presented by Holmes Rolston in his excellent essay, "Valuing Wildlands," *Environmental Ethics* 7 (1985). The present form of this chapter owes much to Rolston's paper and to helpful comments by him on an earlier draft. For a more general discussion of wilderness values in the United States, see Roderick Nash, *Wilderness and the American Mind*, 3d ed. (New Haven: Yale University Press, 1982).

[3] Hammack and Brown, *Waterfowl and Wetlands*, p. 4. The difference between actual benefits received and expenditures is referred to as "consumer surplus."

ing to pay for experiences of natural habitats and species, or to determine how much compensation they would be willing to accept in order to forgo such experiences voluntarily.[4]

Contingent valuation has some promise in assessing recreational options.[5] Many types of recreation are purchased in markets, so there are useful analogies available to guide respondents in naming a figure. For example, the value of one day of wilderness recreation can be compared to the amount a family would spend in a day at Disneyworld, with proper allowance for differences of taste and differences in costs of providing the recreational service. A number of experiments have, accordingly, been done to place contingent values on recreational opportunities.[6] For example, D. Brookshire and associates administered a questionnaire to hunters in Laramie, Wyoming, to determine their willingness to pay for "encounters" with various wildlife. They found that respondents were willing to pay fifty-four dollars per year to increase encounters from one to five per day.[7] But one would expect considerable variation across species, as elk hunters presumably value encounters with nongame species less than further sightings of elk.[8] These figures have a tempting solidness to them but it is unclear how they could ever support a reasonable policy of preservation because, unbeknownst to respondents and even experts, some highly valued species may depend for their continued existence on a lowly valued species and it is unclear how such valuations could adequately reflect such interdependencies.[9]

Advocates of contingent valuation of recreational activities

[4] See ibid., pp. 6ff., for a detailed discussion of these techniques. For a critical evaluation of them, see Rolston, "Valuing Wildlands," pp. 31-38.

[5] Rolston, "Valuing Wildlands," p. 32.

[6] For an overview, see William D. Schulze, Ralph C. d'Arge, and David S. Brookshire, "Valuing Environmental Commodities," *Land Economics* 57 (1981): 151-172.

[7] D. Brookshire and A. Randall et al., *Methodological Experiments in Valuing Wildlife Resources: Phase I Interim Report to the United States Fish and Wildlife Service*, 1977.

[8] Ibid.

[9] See Stephen R. Kellert, "Social and Perceptual Factors in the Preservation of Animal Species," in *The Preservation of Species*, ed. Bryan G. Norton (Princeton, N.J.: Princeton University Press, 1986). Also see Chapter 13, below.

recognize that dollar figures mentioned by respondents may not accurately reflect true values. For example, they worry that respondents who guess the purpose of a survey may inflate figures they are actually willing to pay (in order to protect their favorite recreational area) or deflate them (in order to hold down users fees).[10] Designers of surveys therefore build controls against such biases into their questionnaires, allowing for "correction" of these biases.[11] By including control questions, they can adjust responses to reflect more adequately what the person would actually pay *if there were a fair market for the recreational activity in question.* Biases such as these, which might be called "intentional" biases, can no doubt be minimized, but few would attribute to the resulting estimates more than approximate validity.

Critics of contingent valuations, such as Holmes Rolston, III, however, believe that there is a deeper, "unintentional" bias built into the framing of questions.[12] Rolston would argue that even if responses are corrected for intentional biases and the figure assigned represents what users would actually pay if a fair market existed for wilderness recreation, such dollar figures do not accurately reflect the true value of wilderness activities: "Such categories as existence, option, and bequest values promise to package up a fuzzy assortment [of values], but as values grow intangible, social and ecosystematic, the individual's capacity to price them becomes progressively poorer.[13] In addition, Rolston points out that recreation is, after all, the best case for valuing natural amenities because of the fairly straightforward market analogies that exist for other recreational opportunities. It is much more difficult to conceptualize markets, actual or hypothetical, for other amenity values such as aesthetic value, scientific understanding, and religious and symbolic values of wild

[10] Rolston, "Valuing Wildlands," p. 33.

[11] In "Valuing Environmental Commodities," Schulze et al., discuss "strategic bias," "information bias," "instrument bias," and "hypothetical bias" (pp. 155-159). These are all biases that cause respondents to state dollar figures that do not correctly reflect their true willingness to pay or their willingness to accept.

[12] This term is mine. I use it as a shorthand to refer to inaccuracies that would remain even after intentionally misleading assessments have been corrected for.

[13] Rolston, "Valuing Wildlands," p. 35.

species and natural places. In these cases it can be argued that special features of aesthetic and related experiences of nature make unintentional biases likely. That is, even earnestly stated figures of willingness to pay or accept may not reflect the true value of the experiences in question.

In the remainder of this chapter I will discuss the special features of amenity values that make accurate contingent valuations particularly difficult. In effect this discussion will lead us through a catalogue of some unique aspects of natural aesthetic and other amenity experiences. These include: (1) the difficulty of isolating aesthetic aspects of a naturalistic encounter from broader experience; (2) the dynamic or transformative value of symbolic, religious, and aesthetic experiences of nature—their special contribution to transformative values; and (3) the "subjective" or "relativistic" nature of aesthetic and other amenity values derived from nature.

5.3 The Pervasiveness of Amenity Values

Attempts to assign contingent values to recreational opportunities have some plausibility because it is possible to view recreation as consisting of a collection of relatively discrete activities. Accordingly, the annual value that an individual places on an outdoor recreational area might be represented as the dollars she would be willing to pay to use that area for one day, multiplied by the number of days she actually uses it in a year. Using this method, one might even arrive at dollar values assigned to particular species if, for example, the amount she would be willing to pay would decrease by a specifiable amount if the last grizzly bears were extirpated from the area.[14] Suppose, however, that the task is to specify a contingent valuation for the symbolic value of the bald eagle. Now it is more difficult to focus on any particular unit of experience. When and where does one "symbolically value" the eagle? In the case of recreation, estimates could be

[14] See J. Baird Callicott, "The Land Aesthetic," *Orion* 3 (1984): 23 [reprinted from *Environmental Review* 7 (1983): 345-358], for a discussion of Aldo Leopold's views on "ecological indicator species" and their aesthetic importance.

tested experimentally by systematically raising user fees for an area and noting at what point usage declines. But there is no analogous means to limit access to the symbolic use of eagles. Here, perhaps, willingness-to-accept tests would be more appropriate. Respondents could be asked, "How much compensation would you require to accept willingly the extinction of the eagle?" But this strategy pushes the respondent onto unfamiliar territory as a seller of symbolic values and no guidance is forthcoming from actual or mental experiments with user fees.

Suppose our respondent courageously assigns a dollar value (fifty dollars per year, say) to accept the extinction of the eagle. What exactly does this figure represent? One way to answer this question would be to imagine under what conditions the respondent might correct or regret her answer. Suppose she writes down "fifty dollars per year" on a questionnaire, and then suppose that policy choices leading to the extinction of eagles are subsequently made. From the date of the extinction our respondent receives a check for fifty dollars each year for the remainder of her life. Unlike the case of hiking in the woods before and after the extirpation of grizzlies, where the respondent might note that she stops going to that area in spite of lowered user fees after the extirpation, our respondent finds it difficult to specify the ways the loss of the eagle affect her. She may feel less patriotic; she may even be less likely to fly the flag on national holidays. But these changes are very difficult to price. Further, they do not come close to exhausting the full range of changes in her experiences that result from the loss of the national symbol; the entire experience of "being an American" is subtly altered.[15] But what is this experience worth to her in dollars—at what point would she say that she wishes she had responded with a larger figure, such as one hundred dollars per year?

One reason symbolic and aesthetic values are difficult to price is that they are diffused through experience—they often cannot

---

[15] See Nash, *Wilderness*, and Mark Sagoff, "On Preserving the Natural Environment," *Yale Law Journal* 84 (1974): 205-267, for discussions of the particular importance of wild species and natural ecosystems in the formation of American values.

be associated with specific experiences but rather affect and infect the quality of experience as a whole. Cases such as putting twenty-five cents in a telescope dispensing three minutes of view are unusual and unrepresentative of amenity values. The difficulties in monetarizing these amenity values is in some ways analogous to the problems, explained in previous chapters, of assigning values to particular species in isolation. By posing the species-valuing question as one of listing specifiable products and services derivable from a particular species, benefit-cost analysts tend to ignore the contributory value of species. By contributing to the support of other species in unknown and unspecifiable ways, species that have no known uses contribute indirectly to the provision of commodities that are provided directly by species they support through ecological interactions. A similar line of reasoning applies to amenity values.[16]

If one attempts to assign a dollar value to symbolic or aesthetic experiences connected with a particular species or particular experience, one risks ignoring crucial interactions among these units. Some species that have little or no amenity value in themselves may support other species that do, and consequently they have "indirect" aesthetic value. More importantly, the concentration on discrete units of aesthetic value obscures the interanimation of aesthetic experience. Diffused through experience, these amenity values affect the quality of enjoyment, not their quantity or duration.

These arguments concerning aesthetic value also present an interesting parallel to the argument that mature ecosystems can make a valuable contribution to total diversity without being themselves the most diverse.[17] The strength of aesthetic arguments, in a parallel way, rests upon the total diversity of human experience. One need not claim that experience of nature surpasses experience of art in either its diversity or its intensity in order to argue for its contribution to the diversity and intensity of human experience generally. Just as diversity of the mosaic intro-

[16] See Section 3.4.
[17] See Section 4.3.

duces new ranges of biological diversity, exposure to it introduces new ranges of human experience. Aesthetic experience of art does not compete with aesthetic experience of nature. Each heightens and deepens the other.

Large urban areas are often considered attractive areas in which to live because of ethnic, cultural, and other opportunities they offer. On one level this diversity is valued because it presents options, considered discretely. If one can attend an ethnic festival on Saturday afternoon, a symphony on Saturday evening, and a scientific lecture on Sunday afternoon, the diverse offerings satisfy a whole range of felt preferences. But these experiences viewed discretely represent only the beginning of the value of such diversity because the juxtaposition of such opportunities creates a cross-fertilization of ideas and attitudes. Experiences do not happen discretely but build upon and illuminate each other. Likewise for aesthetic experiences of art and nature. If aesthetic experiences are viewed as having so many units of value, assessed one by one, it might be argued that a Disney World of artificial environments would suffice. With enough ingenuity, units of aesthetic experience of a given intensity may be made to order. But aesthetic experiences of wild species and undisturbed ecosystems cross-fertilize experiences of art and artifacts, and each enriches the other.

In a discussion of demand values assigned to other species it is natural to speak of aesthetic experience as a commodity. There is no doubt that it can be such, as in the case of the telescope at the scenic lookout. In this and similar cases the problems involved in computing a monetary value for such discrete aesthetic experiences may be purely technical. But when one recognizes the way in which varied aesthetic experiences illuminate and change one another—a hike in the mountains heightens the experience of a painting by Frederic Church, or vice versa—total value exceeds the summed value of individual experiences. This pattern is a pervasive one, and it explains the inability of the above-described respondent to feel confidence in assigning a dollar value to the symbolic experience of the bald eagle. Concentration on particular discrete events, whether species extinctions or aesthetic experiences, ignores the contributory and cross-fertilization values that are part of a diverse and complex system. These holistic values emerging from complex systems composed of discrete parts are the essence of "community values." They cannot be measured and monetarized individually without significantly narrowing and falsifying their true worth.

Because of differences inherent in the experience of natural, as opposed to artistic, aesthetic objects, these holistic, diffuse qualities of aesthetic experience are especially pronounced in the natural case. Ronald Hepburn has examined three ways in which aesthetic experiences of nature differ from aesthetic experiences of art objects.

(1) Aesthetic experiences of art objects normally involve a detachment, a "standing-back-from" the object. While aesthetic awareness of nature can involve the detachment of the observer from the object, it more often envelops him on all sides. Experiences of nature often involve immersion in a total environment, a complete experience of all senses seldom attempted in the presentation of art objects.[18]

(2) Art objects often have frames or pedestals designed to set them apart from their environment. Natural objects are frameless and, consequently, the focus of concentration is more variable as the same object may at times be viewed individually and at other times as part of a changing landscape, the "frame" is supplied by the perceiver in the perceptual act of appreciating natural beauty. This different process provides more latitude for perceptual variation interpersonally and for intrapersonal variation across time.[19]

(3) Because natural objects are not created intentionally as aesthetic objects, they offer great scope for individual imagination. We can attribute shapes to cloud forms, or enjoy a beautiful butterfly without wondering whether we got the artists's intentions "right." In viewing an art object we take satisfaction in grasping

[18] Ronald W. Hepburn, "Aesthetic Appreciation of Nature," in *Aesthetics in the Modern World*, ed. Harold Osborne (New York: Weybright and Talley, 1968), p. 51.
[19] Ibid., pp. 51-52.

its intelligibility as a perceptual whole, based on guides to interpretation provided by the artist. No such guides exist in perceiving nature, so a freer form of interpretation naturally occurs.[20]
Aesthetic appreciation of nature, then, has special characteristics
that are seldom present in the appreciation of artifacts, characteristics that enhance those values that are not appreciated in discrete chunks.

When we think of aesthetic objects, we often think of beautiful
artifacts or scenic vistas. But this illegitimately constricts the
scope of the aesthetic. Some of the greatest paintings are better
described as jarring, shocking, even ugly, than as beautiful. Sagoff observes that the value of Picasso's *Guernica* and Van
Gogh's *Rooks Over a Cornfield* lies not in being beautiful or dec-
orative but in being profound, expressive, and passionate.[21]

Much the same holds true of aesthetic objects in nature. Many
species provide aesthetic experiences but not by being beautiful.
Using a somewhat wider sense of the term "beauty," the Ehrlichs
distinguish between "conventional beauty" and the "beauty of
interest," including under the latter category many things not
normally called beautiful at all, such as many insects and other
invertebrates.[22] Yet when people examine such species more
closely, they can be amazed, fascinated, and delighted at the in-
tricacy and the ingenuity of the adaptations involved.

It would be a mistake, then, to equate aesthetic experiences of
nature with experiences of conventional beauty. Nor must all aesthetic experiences be pleasurable. One can be awed, shocked, or
fascinated by the experience of a predator killing and devouring
its prey, even though it would be inappropriate to describe the experience as pleasurable. The moving or the shocking is as much a
part of aesthetic experience as an experience of beauty.

Aesthetic experience of nature, as well as aesthetic experience
of artifacts, comprehends a tremendously broad range of experience. Loss of a species of great aesthetic value shares a number of

[20] Ibid., p. 53.
[21] Sagoff, "On Preserving the Natural Environment," p. 210.
[22] Paul Ehrlich and Anne Ehrlich, *Extinction: The Causes and Consequences of
the Disappearance of Species* (New York: Random House, 1981), pp. 38-39.

characteristics with destruction of a great work of art.[23] But aesthetic experience of natural objects has special characteristics not normally found in experience of artifacts, and these similarities and differences interanimate experience and create diversity and richness that is diffused throughout experience and not easily measured. Loss of a great natural symbol, such as the bald eagle, a beautiful species such as the whooping crane, or an unusual one such as the manatee detracts from the quality of human experience as a whole. In this sense contingent valuations, even when they eliminate the intentional biases that occur when respondents misrepresent what they would actually pay for recreational and other amenity values, may be biased in a deeper, unintentional sense. The dollar values specified seem unlikely to take full account of the diffuse and qualitative effects of amenities on the totality of experience. The framing of the question as one of assessing the willingness to pay for a discrete experience discourages the respondent from including the diffuse, interactive value of amenity experiences, much as asking a respondent to place a value on the discrete periods of time spent with a loved one would hardly lead to a full valuation of the relationship.

I began this section by agreeing with Rolston's assessment that, among amenity values, quantitative economists should have the easiest time with recreation values. But the arguments just developed apply likewise to recreation, if one recognizes that recreation, to, consists of more than discrete events such as a walk in the woods or a "user-day" at a wilderness area. If a walk in unspoiled mountains enhances a viewer's next encounter with a landscape painting and experiences of nature illuminate human social encounters, it will be misleading to ask respondents to evaluate recreational use of a natural area in terms of daily user fees.

5.4 The Transformative Nature of Aesthetic Experience

For convenience in computing demand values it is necessary to treat the felt preferences on which they are based as fixed. But this

---

[23] See Elliott Sober, "Philosophical Problems for Environmentalism," in *The Preservation of Species*.

is only a methodological necessity—in fact, each person experiences a complex and shifting pattern of preferences and priorities. This pattern is, at any given point in time, shaped by genetic, emotive, and cognitive forces affecting the individual. In discussing the source of amenity values we focus on the latter two forces, emphasizing the development of a pattern of nonessential desires and tastes that lend richness and meaning to human experience and are not determined by the genetic requirements of existence.

Because they are by definition based on nonessential needs, aesthetic and other amenity values are particularly dynamic in nature—they change and develop as humans mature and expand their experience. Exposure to art and fine architecture, for example, lead to increasing sophistication. As a young person views and learns about art, these new experiences inform and alter preferences. What seemed exciting and satisfying to the novitiate becomes coarse and unfinished with wider experience. Similar processes occur with exposure to nature. As a young child accompanies his parents on nature hikes, he learns about the local flora and fauna. On progressively longer walks, he is exposed to new challenges, to ascending steeper and more rugged slopes. Gaining in proficiency, he may strike off on his own, developing an appetite for still greater challenges. These, in turn, increase self-confidence and expand horizons as the child begins to prefer wilder and more inaccessible places and gains a taste for solitude and for tests of individual strength and self-sufficiency.

When aesthetic and other amenity values are discussed as merely demand values, as reflections of artificially fixed felt preferences, this dynamic aspect is underemphasized. Environmentalists often argue that exposure to natural places and wild species has beneficial consequences for human perception and leads to improvement in the quality of human lives. These benefits, which depend upon alterations in felt preferences and which I have called transformative values, go beyond, and cannot be properly understood within, the limited framework of fixed demand values, a point I will expand in Part C below.

Here I want only to call attention to one special feature of such arguments and their relationship to natural aesthetics. I have defined transformative values neutrally: if exposure to great art causes a positive transformation in human perception and understanding, it is also possible that exposure to bad, trashy art will have a corresponding negative effect. Species preservationists who employ arguments based on transformative values confidently treat exposure to wild species and natural places as having *positive* value. What justifies this confidence? Why are they not vulnerable to a counterargument that experiences of nature, of predators killing prey, for example, have negative transformative value and that species who violently attack other species should be systematically eliminated to protect the morals of human society?

At least an important part of an answer to these questions is provided by the thesis, widely held among environmentalists and natural aestheticians, that nature has only positive aesthetic qualities. Allen Carlson states this view as follows:

> [T]he natural environment, insofar as it is untouched by man, has mainly positive aesthetic qualities; it is, for example, graceful, delicate, intense, unified, and orderly, rather than bland, dull, insipid, incoherent, and chaotic. All virgin nature, in short, is essentially aesthetically good. The appropriate or correct aesthetic appreciation of the natural world is basically positive and negative aesthetic judgments have little or no place.[24]

Admitting that this view is initially implausible because it implies an important disanalogy to aesthetic appreciation of art, which essentially involves negative as well as positive judgments, Carlson goes on to show how widely it is accepted among natural aestheticians.[25] He considers and rejects several arguments that purportedly establish the thesis that nature has only positive aesthetic qualities and then argues that the view can be justified if

[24] Allen Carlson, "Nature and Positive Aesthetics," *Environmental Ethics* 6 (1984): p. 5.
[25] Ibid., pp. 5-10.

one notes the "intimate connection between nature appreciation and the development of natural science."[26]

Carlson recognizes that this form of justification for positive natural aesthetics imposes an important qualification on the view.

> The justification [the natural aesthetician] develops regards aesthetic appreciation of nature as informed and supported by the development of science. A positive aesthetics of nature seems, therefore, to depend upon the scientific world view. The positive natural aesthetics that is justified thus depends on the interpretation of scientific knowledge. If science is viewed, for example, as absolute, positive aesthetics will hold absolutely. If, on the other hand, science is viewed as culturally relative, positive aesthetics should be viewed as culturally relative as well.[27]

Given that environmentalists generally accept the scientific, naturalistic world view, this qualification will not bother them. Beliefs they have already accepted encourage them to use positive natural aesthetics to support arguments from transformational value.

Carlson notes that, historically, aesthetic appreciation has followed advances in scientific understanding:

> The positive aesthetic appreciation of previously abhorred landscapes, such as mountains and jungles, seems to have followed developments in geology and geography. Likewise, the positive aesthetic appreciation of previously abhorred life forms such as insects and reptiles, seems to have followed developments in biology. In retrospect, many of the advances in natural science can be viewed as heralding a corresponding advance in positive aesthetics.[28]

These points are developed cogently in Baird Callicott's essay, "The Land Aesthetic." Callicott attributes to Aldo Leopold a

[26] Ibid., p. 5.
[27] Ibid., p. 32.
[28] Ibid., p. 33.

land aesthetic, parallel and complementary to his more familiar land ethic. Both are situated squarely upon scientific foundations of ecology and evolutionary biology.[29] Experience is then informed by the categories and theoretical constructs of biological science:

> There exists a subtle interplay between conceptual schemata and sensuous experience. Experience informs thought. That is true and obvious to everyone. What is not so immediately apparent is that thought equally and reciprocally informs experience. The "world" as we drink it through our senses, is first filtered, structured, and arranged by the conceptual framework or cognitive set we bring to it, prior, not necessarily to all, but to any articulate experience.[30]

When Callicott emphasizes the reciprocal relationship between experience and cognitive knowledge, he integrates the crucial aspects of the preservationists' argument from transformative values. Perception changes the cognitive structure of our world view and, within this structure, experience gains depth and breadth: "Ecology, as Leopold pictures it, is the biological science that runs at right angles to evolutionary biology. Evolution lends to perception a certain depth, 'that incredible sweep of millennia,' while ecology provides it with breadth. Wild things do not exist in isolation from one another."[31] Thus, within the ecological world view, science and aesthetic perception inform and alter each other, as species are seen to perform important community functions and the entire pattern of relationships is perceived as existing in interlocking, meaningful community relationships.

When aesthetic experience is understood in its deeper dimensions, informed by ecological and evolutionary science, it is necessarily positive. Seen in this light it is also necessarily dynamic: aesthetic understanding also acts to alter and change emotional reactions to objects and affects the values and preferences that are

[29] Callicott, "Land Aesthetic," p. 19.
[30] Ibid., p. 20.
[31] Ibid., p. 19.

felt and expressed.[32] Thus, while aesthetic values have an important demand dimension—they fulfill human preferences as felt—they also have an even more important dynamic dimension, altering and transforming those same felt preferences through time. To insist that aesthetic and other amenity values must be monetarized, reduced to expressions of felt preferences in actual or hypothetical markets, is to freeze them illegitimately in time for purely methodological reasons. It also ignores the very important contribution of experiences of wild species and natural ecosystems to transformative values.

5.5 Subjectivity and Aesthetic Value

I have emphasized the ways in which such special features of amenity values as aesthetic and symbolic experience may cause policy analysts to miss important aspects of their value. Another feature of these experiences, their subjectivity (or, alternatively, their relativity), is sometimes thought to justify discounting the importance of these values. Arguments involving aesthetic and amenity values are often given less credence in discussions of species and wilderness preservation because they are more dependent, in some sense, on the viewpoint of the observer. Is there a sound argument that begins with the premise that the aesthetic value of nature is subjective or relative and concludes that aesthetic arguments must be given diminished weight? One argument of this sort begins by asking whether aesthetic qualities of nature are "subjective" or "objective." This issue is then understood to depend on whether aesthetic qualities are qualities of the observer or of the observed object. It has been argued that all aesthetic qualities, whether of nature or of art, are subjective and, hence, arbitrary; it has also been argued that aesthetic experience of nature is more arbitrary than is experience of art.[33]

---

[32] See Carlson's discussion of how scientific "location" of a species or event is analogous to categorizing a work of art correctly within its appropriate genre ("Nature and Positive Aesthetics," pp. 28-34).

[33] See Kendall Walton, "Categories of Art," *Philosophical Review* 79 (1970): 334-367. Also see L. Duane Willard, "On Preserving Nature's Aesthetic Fea-

The debate about subjectivity affects the question of whether aesthetic values provide a reason for preserving species only if the subjectivity in question undermines the value of an experience. But I see no reason to believe that it does so—experiences can be equally valued, whether they are objective or subjective. Whether aesthetic qualities are objective or subjective, it is admitted on all hands that the value of the aesthetic experience depends on its effect on human valuers. And it is this question of value, not the locus of aesthetic qualities, that is the present concern. Even the most vigorous subjectivist does not deny that great art has value. So, if species are given value of the same type as art, this will be sufficient to support a rationale for preservation, provided humans are likely to encounter and enjoy them.

A valuable aesthetic experience clearly requires both a perceiver and an object perceived. A valuable aesthetic experience of other species requires, in addition to a valuer, that there be some specimens of the species in question available for human appreciation. The continued existence of some specimens of a species, the avoidance of extinction, is a necessary condition for deriving aesthetic value, regardless of whether the aesthetic qualities of value are considered qualities of those specimens or qualities of the perceiver.[34]

Another argument concludes that aesthetic value should be given less weight on the grounds that it is relative to cultural background. Subjectivism may seem to support relativism. If aesthetic experience of nature depends mainly on persons and not on the qualities of natural objects, then it may be more plausible to say that this value depends on the cultural background of experience.

But relativism of *value* is ultimately independent of subjectivism in the *locus* of aesthetic qualities, and it is the value of the aes-

---

tures," *Environmental Ethics* 2 (1980): 293-310, for a detailed analysis of these arguments.

[34] It may be true that humans can derive aesthetic value from extinct species by reading about them in books and by reconstructing their physiology and ecology through the fossil record. But nobody doubts that these experiences would at least be enhanced by having living examples of the species as well.

thetic experience with which I am concerned in this discussion. The relativism of values said to characterize aesthetic qualities could occur even if these qualities are qualities of objects, not persons. Even if aesthetic qualities are qualities of objects, the values attached to those qualities could be relative to culture. Two cultures, differing in history and value constructs, could agree that a particular landscape is powerful while differing on whether it is an expression of God's power or an expression of the forces of evil.[35] Thus, it would be possible to be a cultural relativist about the *value* of aesthetic experiences, whether one believed aesthetic properties to have their locus in the objects (objectivism) or in the perceiver (subjectivism).

In any case, it might be argued that aesthetic values should be given less weight because they are relative to culture. Let us assume the controversial claim that aesthetic values are relative in this way. What, exactly, would follow? We saw previously that aesthetic response is not limited to judgments of beauty. Some aesthetic value is assigned by all cultures to interactions with nature. If wild species are judged threatening in a given culture, they are still powerful expressions of that culture's world view. They play an expressive role within that culture, and the responses they elicit are properly called aesthetic, whether they are responses of awe or terror. If a tribe has a ritual in which the elders dress as wolves and dance about a fire, the experience of participants and observers can be aesthetic whether the wolf is emulated as an example of freedom and wisdom or symbolically slain as a representative of evil threatening tribal fortunes. In either case the wolf is valuable as an expression of important cultural attitudes.[36]

That such reactions are relative to culture does not imply that they are without value or of lesser value. Insofar as we are here viewing aesthetic experiences as satisfying demand values, they

[35] This seems a consequence of Sagoff's relating the *American* conception of freedom and independence to factors determined in American history. See Sagoff, "On Preserving the Natural Environment."

[36] Here Carlson's qualification (quoted above at note 25) is relevant: cultures that do not ascribe to the scientific world view might perceive negative aesthetic qualities in nature.

provide experiences that fulfill particular felt preferences. To the extent that such symbolic experiences *are* sought, they necessarily have demand value. It does not matter that the particular content or quality of those experiences vary across cultures; what matters is that some such experiences are sought in most cultures. And they are often sought at great cost to other, nonaesthetic values, as when the best deer is sacrificed to the god of the hunt. That varied cultures have differing customs and associate differing aesthetic and cultural values with nature and wild species does not affect the undeniable fact that nature and wild species do provide pleasurable and/or symbolic experiences. To the extent that such experiences are sought, often at great cost to more practical, commodity values, they are valued within the culture that seeks them.

Aesthetic and other amenity values thrive on diversity. Humans seek varied experiences to avoid boredom but also because variety promotes cross-fertilization of ideas and attitudes. Even if aesthetic experiences of wild species are more subjective and relative than are more "practical" uses of nature (and these points remain controversial, of course), it does not follow that they should be given diminished value. Diversity of experience within cultures and across cultures is valued in itself. Aesthetic experiences of nature and wild species supplement and complement the aesthetic values attached to human art objects. The experience of nature contributes to diversity of experience generally.

Aesthetic experience and other amenity uses of wild species, then, provide an important category of demand values. Even though varied cultures experience nature differently, it is universally true that the life of any culture will be richer and more diverse if wild species are encountered aesthetically and in recreational enjoyment, not just as resources for commodities. These experiences involving wild species provide yet more reasons justifying the preservation of species.

Arguments concerning nonessential or amenity values are often given limited credence in discussions of species preservation. Besides the reasons mentioned in this section, there seems to be a prejudice against them simply because of their nonessential character. Essential considerations are felt to be preemptive in determining land use: starving humans cannot enjoy beautiful views. To this extent aesthetic experience must fit into hierarchies of value in a different way from survival needs. But I have shown how the priority of survival needs in this context is balanced against a very different sort of priority that aesthetic experience can claim. The quality of our experiences, the amenities of aesthetic enjoyment, scientific understanding, symbolic and religious values supported by wild species, can shape the very demands that ultimately claim priority over them. Amenity values, because of their dynamic nature and transformative powers, are too rich to be reduced to dollars in markets, whether these are actually paid or hypothetically inferred.

S I X

A PARTING LOOK AT
DEMAND VALUES

6.1 Human Benefits and Economic Analysis: A Summary

I began Part A of this book by arguing that the way in which the question of the value of a species is framed can lead to quite different answers to that question. One approach examines species individually, attempting to identify particular industrial, commercial, option, and existence values of species considered one by one. This approach is favored by economists who advocate a BCA approach (or various modifications of it) to decision making regarding species. It is an essential assumption of their approach that meaningful, nonarbitrary dollar values can be assigned species in a way that does not systematically bias the resulting decisions.

In contrast to this approach, I suggested that one might assign values to species in more general terms, beginning by trying to ascertain the extent to which each and every species has a prima facie value, independent of its populational and individual characteristics. Once such a value were established, it would be possible to examine individual species for particular values that could be added to it. There is no assumption here that these values can be assigned nonarbitrary dollar figures. This framing of the question, then, complements the SMS approach to decision making regarding species. According to that approach, it is urged that species be saved, provided the costs are not intolerably high. In essence, the SMS approach assumes considerable positive value for all species and abandons the attempt, central to BCA approaches, to assign values to the benefits side of the account. The problems of the SMS approach are: (1) the policies it recommends are only as defensible as its assumption that all species are of considerable value, and (2) its central criterion is vague, relying as it does on the unoperationalized phrase "intolerably high social costs."

One goal of the first part of this book has been to evaluate the strengths and weaknesses of these two approaches to economic decision making. I believe that the arguments detailed in these chapters show that one of the central assumptions of the BCA approach, that meaningful and nonarbitrary dollar values can be assigned individual species, is false. Even in cases of species with well-known commercial values, it is unclear how to compute values for secondary uses that might be consistent with current uses. Even if current demands for access to useful assets reflect the expectation that new uses will be discovered, these added uses are likely to be discounted for risk aversion and time preference. Thus, even assignments based on clearly understood, present commercial uses should be augmented by less easily evaluated uses that are still undiscovered.

When one tries to ascertain the dollar values of option values, quasi-option values, and existence values, debilitating problems arise. Ecologists do not understand ecosystem functioning well enough to assign values to systems, let alone to individual species within these systems. While decisions are usually made in situations where complete information is lacking, the ignorance in the present case is pervasive and compounded by uncertainties on two distinct levels. To force the assignment of dollar values as demanded by the BCA methodology is to create a false sense of confidence in unreliable information. Finally, the very attempt to assign values to species individually prejudices the case against species by obscuring and ignoring the general, prima facie value that every species has. These arguments demonstrate the great danger that species will be systematically undervalued on the BCA approach.

Further, some of the same arguments that undermine the central assumption of the BCA approach lend credence to the central assumption of the SMS approach. Five arguments discussed in

Chapter III and Chapter IV uncovered powerful reasons to save every species. Because species constitute total diversity and contribute causally to the creative function of that diversity, two arguments show that the extinction of a species must be seen as a real loss. First, any species that is lost will no longer contribute to the ecological and evolutionary processes that create specialized adaptations and, second, a species that is lost may lead to a cascading wave of extinctions. Both the potential species that will not emerge and the existing species that will be extinguished will surely include some that would have proved useful to humans. If these two arguments might be thought to show only a minimal prima facie value for each species, further arguments strengthen and enhance that value. Since the biological diversity of the planet has already entered an accelerating downward spiral, losses of species represent further accelerations toward local and global ecosystem breakdowns. The risks of such breakdowns are so great and the contribution of species losses to them are so little understood that any rational society would exercise extreme caution in contributing to that acceleration. In addition, the processes that create new adaptations, interrelationships, and species also create mature autogenic systems. These dynamically stable systems are characterized by efficient and closed food webs and consequently contribute to the regeneration of overexploited resources. Finally, the species most likely to be threatened turn out to be the species most likely to be useful to humans. Therefore, the prima facie value established by the first four arguments is increased by a considerable factor deriving from the correlation between the susceptibility of species to extinction and their usefulness to humans.

These arguments undermine the central assumption of the BCA approach and simultaneously support that of the SMS approach. The values of species are shown to be considerable but not realistically quantifiable. These arguments also flesh out the criterion of the SMS approach. Judgments using the vague phrase "intolerably high costs" are now seen to be informed by multiple arguments showing substantial value for each species. Further, this value is ascending—every species loss contributes to a descending spiral of biological diversity and makes remaining species all the more valuable. While the SMS approach cannot rival the promise of the BCA approach to provide a quantified dollar figure to set against the benefits of development activities, it does provide strong, if unquantified, guidance to replace that unfulfilled (and unfulfillable) promise.

Still the skeptic might ask, how is the SMS criterion to be employed in making difficult decisions? The answer is as follows. While there is no quantifiable criterion, the arguments support a general policy against allowing further extinctions. That is, the case is strong enough to shift the burden of proof from species preservationists to developers. The arguments yield only prima facie values, values that could be overridden. But if a sufficiently long view is taken of human benefits, sufficient to avoid undermining the health and well-being of future generations, then mere convenience or short-term economic advantage would not justify the decision to cause or allow another species extinction.

Perhaps an analogy would be useful here. Consider the decision of a confirmed alcoholic whether to have a drink after learning he has suffered irreversible liver damage. He can be viewed as facing a trade-off between the advantages of having the drink and the undoubtedly minimal liver damage to be expected from one more drink. So conceptualized, the decision will be made different ways in different cases, depending on the benefits perceived to accrue from taking the drink. But since it is virtually impossible for a confirmed alcoholic to have just one drink, medical experts counsel that the alcoholic adopt a policy against ever taking a drink. To treat the decision whether to have a drink in particular situations without considering the impetus it creates toward further such effects is to prejudice the case in a disastrous manner.

The comparison should be obvious. The scientific evidence that diminutions in diversity compound themselves through time is analogous to the medical evidence that alcoholics cannot take one drink. To treat the decision to allow a species to go extinct as an isolated choice between losing a single species and gaining some short-term economic benefits is to misapprehend the issue. A forward-looking society should have a policy against decreasing biological diversity, as the alcoholic should have a policy against drinking. Both policies are, in a sense, based on prima facie values—they could be overridden. No doctor would tell an alcoholic to die of thirst when no nonalcoholic beverages were available. But the doctor would surely recommend to the alcoholic that he avoid such difficult situations at considerable cost and that truly extraordinary conditions must obtain before the act of taking a drink becomes rational. This analogy suggests that almost all costs of preserving a species should be considered "reasonable," thereby giving substance to and strengthening the SMS criterion and supporting a general policy of species preservation.

6.2 The Inadequacy of Demand Values

I have also tried to make as strong a case as possible for the claim that every species makes an important contribution to human demand values. Beyond all the standard and familiar ways in which nonhuman species contribute to human welfare, I have pointed out powerful additional contributions that are easily overlooked and impossible to quantify.

Human demand values, the reflections of individual felt preferences for certain goods or amenities, appear to be expressible on a single scale of value, a fact that tempts benefit-cost analysts to insist upon dollar figures to represent all such values served by nonhuman species. Where no market exists for the good or amenity in question, benefit-cost analysts fall back upon the WTP and WTA criteria. I have sketched ways in which attempts to assign such dollar figures risk serious and compounded inaccuracies. Sometimes it is best to assert that a species has a substantial value for fulfilling human demands but that information permitting such an assignment is currently unavailable. So the central assumption of the BCA approach is best rejected, even if all values are regarded as demand values.

Yet species preservationists are uneasy with resting their case on the role of nonhuman species in satisfying human demand values alone, and consequently they go on to argue that there are nondemand values that are even less likely to be reflected in dollar figures representing individual felt preferences. These values are not felt preferences at all. Therefore, species preservationists have looked beyond the ways in which species serve human demands in order to emphasize the ways in which they transform human demands. Or, leaving human demands behind altogether, they seek to justify the preservation of species in terms of the intrinsic value that exists in species themselves or in ecosystems. While I will later examine arguments that nonhuman species have intrinsic value and look at the ways in which they alter and transform human values, for the moment it will be useful to rehearse, in detail, the reasons why species preservationists are disinclined to rest their case on demand values alone. I will first examine the charge that appeals to the possible undiscovered uses of species can be dismissed as recommendations for compulsive and often pointless saving. Then I will show how the contingency of demand values can provide at best a shifting and unreliable basis for policy. Finally, I will state what I take to be the most basic reason for preservationists' aversion to purely demand arguments: if human demands are the basis of the threat to species, how can they also support policies protecting against that threat?

6.3 Aunt Tillie's Drawer

The most common argument that invokes demand values to support species preservation appeals to the possibility that any species may someday turn out to be useful in fulfilling some specific human need.[1] Arguments of this type have appeared at several points throughout this part of the book, and they lend powerful support to preservationist policies in general.[2]

[1] See, for example, Anthony C. Fisher, "Economic Analysis and the Extinction of Species," Report No. ERG-WP-81-4 (Berkeley, Calif.: Energy and Resources Group, University of California, November 1981); Norman R. Farnsworth, "The Potential for Plant Extinction in the United States on the Current and Future Availability of Prescription Drugs," paper read at Symposium on Assessing the Economic Value of Plant Species, American Association for the Advancement of Science, Washington, D.C., January 4, 1982; and Norman Myers, *A Wealth of Wild Species: Storehouse for Human Welfare* (Boulder, Colo.: Westview Press, 1983).

[2] See Section 2.1 and arguments 1 and 2 in Section 3.4.

124

But there are limits to how much weight such arguments can bear. One problem, noted in Section 3.5, is that there are so many species and so few researchers available to examine them. Even if a given species might turn out to be valuable *were it examined*, the chances of its being examined soon are very small, and countless other species would have equal chances of yielding benefits if they were examined instead. The argument assumes that species are valuable because they are scarce; but, for the foreseeable future anyway, it is not species but scientific effort that is scarce. Thus one can hardly cite current economic values lost at each extinction. Economic losses would be meaningful only when the stock of unexamined species gets low or includes species particularly likely to yield useful discoveries.

This argument also suffers by analogy to the syndrome of the compulsive collector. Everybody has a favorite eccentric aunt, uncle, or friend who has drawers, closets, garages, attics, and yards filled with objects bought at garage sales or found in dumps and saved "in case I might need them someday."[3] The argument that species should be preserved in case they may prove useful someday exactly mimics the reasoning of such eccentrics. All of us have felt the force of Aunt Tillie's argument when we discover the perfect use for the do-hickey or whatchamacallit discarded yesterday. But we resist that force for the most part. As anyone knows who has had to repaint an automobile because it was parked in front of a full garage, there are costs attendant upon saving junk, and these costs often outweigh the loss involved in pitching out objects with no foreseeable use. This illustrates the weakness of the Aunt Tillie's drawer argument—at the same time that it recognizes a possible, but unspecified, use for a species, it reminds us that there can be economic costs attendant upon saving it. The saving of junk "just in case" is, after all, eccentric and represents a compulsive behavior generalized from relatively few cases where junk has proved worth saving. Unless a use for the species is likely to be discovered soon, it is tempting to opt for the

[3] Hence the label "Aunt Tillie's Drawer Argument," a designation due to Mark Sagoff.

imminent and clearly specified economic benefits of a road, dam, or other development project.

My point here is not, of course, that species have no commercial value or that we should never save them for that reason. It is only that the argument based upon possible usefulness has drawbacks as well as advantages. If species are valued merely for their possible usefulness, then they become commodities like any other and must compete for space and scarce resources with other commodities. Because there are so many species, it will often seem as though a dam, road, or development project can, within the limited time periods of normal economic analyses, win out in competition. The considerable advantages of preserving species will thus be undermined "incrementally." At many particular decision points the economic advantages of development may exceed the probable values of preservation.[4] But after many species are lost and with them their contributory value, the benefits sacrificed may greatly exceed the total gain of the development projects and a tragedy will have been justified piecemeal. Thus, the Aunt Tillie's drawer argument is more dangerous than helpful because it encourages the treatment of species, individually, as commodities of at best probabilistic value.

For this reason the argument concerning the contributory value of species, cited in Section 3.2, is more important in supporting species preservation than is the probability that the next species lost will eliminate the chance for an important technological advance. Both arguments depend, ultimately, on the likelihood that lost species may prove useful in the future. Probabilistic reasoning functions quite differently in the two arguments, however. Concentrating on single species, the Aunt Tillie's drawer argument rests on the probability that a given specified species is potentially useful and that its use would have been discovered. The product of these two low probabilities appears very low, especially given that there is a far greater shortage of scientific researchers than of potentially useful species.

---

[4] See Thomas E. Lovejoy, "Species Leave the Ark One by One," in *The Preservation of Species*, ed. Bryan G. Norton (Princeton, N.J.: Princeton University Press, 1986).

The contributory value of species, on the other hand, concentrates on the interdependence of species. Each species lost threatens to begin a cascade of further extinctions, and ultimately diversity itself is threatened. The probability that some species lost in such a cascade will prove useful is much higher than just one species, taken singly, will be. Further, if diversity is eroded sufficiently, there may eventually be a shortage of new and interesting species for researchers to examine. The argument for contributory value, then, is much stronger than the Aunt Tillie's drawer argument because, recognizing the interdependency among species, it computes probabilities for the whole range of likely losses, not just for particular threatened species.

6.4 *The Contingency of Human Demands*

Martin Krieger has asked, "What's Wrong with Plastic Trees?"[5] Krieger's answer to his own question provides a powerful *reductio ad absurdum* of the suggestion that human demand values can provide a basis for the preservation of species and natural areas. Kreiger says: "What's wrong with plastic trees? My guess is that there is very little wrong with them. Much more can be done with plastic trees and the like to give most people the feeling that they are experiencing nature. We will have to realize that the way in which we experience nature is conditioned by our society—which more and more is seen to be receptive to responsible interventions."[6]

He proceeds to observe that natural places and natural objects are often relatively expensive to maintain and difficult to get to. Since human preferences are determined by manipulable reinforcement patterns, it should be possible to replace these luxuries with much less expensive plastic substitutes. If people object, their preferences can be altered by "education."[7]

Much could be, and has been, said in response to Krieger's ar-

[5] Martin Krieger, "What's Wrong with Plastic Trees?" *Science* 179 (1973): 446-455.
[6] Ibid., p. 453.
[7] Ibid.

gument.[8] For our purposes it is useful because it establishes the extent to which the relationship between human demand values and preservation of natural objects is a contingent one. Even if one assumes that, at present, the aggregate of human demand values strongly supports the preservation of all extant species, that assumption is subject to erosion at three crucial points. First, developments in technology can provide new and less expensive artificial means to fulfill human preferences. Krieger's argument illustrates this point.

Second, preferences change. Even if people have strong preferences for protecting wild species and active desires to walk in wild places and encounter species in their natural habitat, these preferences and desires may change. Indeed, one might argue that these are armchair longings for our rustic roots, quite out of place in the modern age. On this view, these preferences will gradually be replaced by more appropriate tastes for high-tech solutions to the problems underlying them. As Krieger points out, preferences can be shaped by education or propaganda, so any preference for natural species, products, or places could be deliberately extinguished if it is thought to be too expensive to fulfill or inappropriate for any other reason. Methods that emphasize demand values only cannot express our sense that such "education" constitutes unacceptable inroads on our freedom.

Third, since the preservation of species, especially of large predators such as the North American timber wolf, can require very large areas of undisturbed habitat, the practice has attendant costs. Increases in the human populations and associated increases in the pressure to use land will drive those costs upward. At some point the costs will exceed the benefits, and it will become rational to modify the preference in question or find some other means to fulfill it.

For these reasons the support currently afforded species pres-

[8] The points of this section have been made persuasively by Mark Sagoff, "On Preserving the Natural Environment," *Yale Law Journal* 84 (1974: 207, and by Eric Katz, "Utilitarianism and Preservation," *Environmental Ethics* 1 (1979): 357-364. Sagoff and Katz direct their arguments against "utilitarian" reasons, but these arguments are equally applicable to my concept of demand values.

ervation by human demand values is contingent and subject to change. Demand values provide an at best shifting and unreliable basis for species preservation.

6.5 Unlimited Human Demands

The reasons thus far cited do not, I think, capture the deepest uneasiness preservationists feel about supporting their case simply by enumerating the ways in which other species fulfill fixed human preferences. A more powerful reason is at work in the following declaration by David Ehrenfeld: "There is no true protection for Nature within the humanistic system—the very idea is a contradiction in terms."[9] Ehrenfeld concludes that nature, and especially other species, must be attributed rights. I will discuss his arguments for this conclusion in Chapter 11, but this brief passage expresses, I think, the attitude of many environmentalists who feel a compelling need to transcend human demand values in citing reasons to protect nonhuman species.

That feeling is based on a correct perception that human demand values are, in the final analysis, the source of the problem. Because humans reproduce too often and consume too much, nature and other species are under attack. Some species are exploited directly; others lose their habitat to the expansion of human monocultural agriculture and forces of urbanization. How can these demand values provide the basis for a solution of the problem they themselves have caused?

Ehrenfeld's concern seems to be that arguments based purely on demand values are unlikely to yield a satisfactory long-term policy for preserving species into the indefinite future. Most people currently have no felt preference for preserving plants and invertebrates and little concern for natural habitats and ecosystems. Of course it is possible to point out subtle and often-overlooked ways in which plants and invertebrates serve felt preferences. But if human felt preferences multiply unchecked, through continued

---

[9] David W. Ehrenfeld, *The Arrogance of Humanism* (Oxford: Oxford University Press, 1978), p. 202.

population growth and increasingly consumptive life styles, then these subtle, diffuse, and long-term advantages in fulfilling felt human needs will be overwhelmed by immediate consumptive needs. A severely overpopulated nation is unlikely to show concern for the long-term health of systems that could be productive in the indefinite future. All space will be required for human dwellings and intensive methods for producing goods and services. Protection of species will become a luxury and finally an impossibility.

This disastrous scenario derives from another sense in which human preferences are contingent: they vary not just in their object but also in their volume. If human population grows unchecked and if each individual in developed societies uses and wastes more products, then the sheer volume of human preferences will be overwhelming. Demands to fulfill even basic human needs will eventually outweigh demands for species preservation—how can it be hoped that hungry humans will prefer nonhuman habitat protection to a good meal?

But this entire scenario rests upon two suppositions: first, that human needs will expand indefinitely, and, second, *given* that basic human needs are unfulfilled, those needs will predominate over preferences for species protection. Once the first assumption is accepted, the latter seems inevitable. A society that condones indefinitely expanding human needs to be met by increasingly limited supplies must eventually be faced with a depressing choice: either fulfill existing basic needs or deny them in favor of protecting nonhuman species. In this stark situation it is not difficult to say what will be sacrificed.

But this dilemma also reveals the crucial weakness of an approach to species preservation that relies solely upon demand values. Unbridled reproduction and conspicuous consumption are themselves behaviors based upon felt preferences. If such felt preferences are merely accepted without criticism, then the result of present trends is predictable—society will face the bleak choice between either denying human needs or eradicating many nonhuman species. The way out of this perplexity is not, of course, to challenge the second assumption, that pressing human needs

130

must be fulfilled. The solution is to rein in felt preferences, including the persistent preferences to overproduce and overconsume.

In the next part I will examine approaches to species preservation that set human demand values against the intrinsic value attributed to nonhuman species. In Part C I will examine another basis for criticizing felt preferences that favor population growth and heavy consumption. I will argue that the transformative value of nature and wild species provide, in themselves, a basis for a more rational value structure.

B

INTRINSIC VALUE AND SPECIES PRESERVATION

S E V E N

ANTHROPOCENTRISM

7.1 Anthropocentrism and Related Theses

Many environmentalists, appalled at the destruction wrought by human consumption of natural products, have attributed intrinsic value to other species, arguing that, however useful they are for human purposes, their full value is not exhausted by those instrumental values. Such attributions deny the thesis of anthropocentrism. That thesis can be stated as follows: only humans are the locus of intrinsic value, and the value of all other objects derives from their contributions to human values.

As was explained in Section 1.2, human values can be interpreted narrowly, to include only human demand values, or more broadly, including transformative values as well. Thus, another dichotomy separates anthropocentrists into two camps. Strong anthropocentrists insist that the value of all nonhuman objects derives from their contribution to human demand values, while weak anthropocentrists recognize that such objects have transformative value as well—they provide the basis for informing and criticizing demand values, as well as fulfilling them. Nonanthropocentrists recognize intrinsic value in the fulfillment of some nonhuman demand values and therefore deny the thesis of anthropocentrism. In this chapter I will examine the positive reasons for defending an anthropocentric value theory. Because denials of strong anthropocentrism can involve the recognition either of nonhuman demand values or of human transformative values, they are therefore ambiguous. I will, throughout this part of the book, interpret anthropocentrism in its weak form. Only

135

then does a denial of anthropocentrism necessarily imply an attribution of intrinsic value to nonhumans.

Before evaluating the reasons that have been advanced to support anthropocentrism, it is worth noting how that view is related to similar theses. Terminologically, "anthropocentrism" is interchangeable with "homocentrism" and "human chauvinism." Sometimes the term "humanism" is also used equivalently, but more caution is necessary in this case.[1] Anthropocentrism as here defined is equivalent to the traditional thesis of Man's Dominion, which, following Richard and Val Routley, can be stated as "the view that the earth and all its nonhuman contents exist or are available for man's benefit and to serve his interests and, hence, that man is entitled to manipulate the world and its systems as he wants, that is, in his interests."[2]

I will evaluate the debate between anthropocentrists and nonanthropocentrists in stages. This chapter inquires whether there are convincing reasons to support anthropocentrism. If conclusive reasons exist, no further examination of nonanthropocentrism would be necessary. Since the reasons usually advanced to support anthropocentrism will prove inconclusive, however, I will devote two further chapters to interpreting and evaluating nonanthropocentrism. Chapter 8 will consider versions of nonanthropocentrism that attribute intrinsic value to nonhuman individuals, while Chapter 9 will examine nonindividualistic versions of the thesis. Both approaches will likewise prove inconclusive—available accounts are either insufficiently clear or insufficiently supported to warrant the development of policies designed to protect species as repositories of intrinsic value.

The failure of nonanthropocentrists to establish the intrinsic value of nonhuman species need not, however, limit species preservationists to human demand values to support their protection-

---

[1] See David Ehrenfeld, *The Arrogance of Humanism* (New York: Oxford University Press, 1981). Ehrenfeld uses the term "humanism" to refer to the assumption that all value and power in the universe derive from human sources. There are, of course, a number of other meanings given this term.

[2] Richard Routley and Val Routley, "Against the Inevitability of Human Chauvinism," in *Ethics and Problems of the 21st Century*, ed. Kenneth Goodpaster and K. M. Sayre (Notre Dame, Ind.: University of Notre Dame Press, 1979), p. 56.

ist policies. Part C will be devoted to transformative values and the development of a coherent and adequate but weakly anthropocentric account of the value of nonhuman species. Thus, while attempts to establish the intrinsic value of nonhuman species prove inconclusive, preservation efforts can be adequately justified independent of such claims.

7.2 The Argument from Consciousness

It is often claimed that only humans can have intrinsic value because only humans are conscious (or *self*-conscious). This argument is, at best, highly elliptical, and it is not at all obvious how the missing details are to be filled in.

One version of the argument begins with the claim that only humans are (self-)conscious and concludes from this that only humans are moral beings, relying on an implicit premise that only (self-)conscious beings can be moral beings. If we add another premise that only moral beings can have intrinsic value, the result would be a valid argument concluding that nonhuman species and their members have no intrinsic value. But while the argument is valid, each of its premises is highly questionable and likely to be challenged by at least some nonanthropocentrists. One would first have to defend a relatively restrictive and contentious definition of consciousness or self-consciousness to support the initial premise.

The implicit premise that only (self-)conscious beings can be moral beings is likewise questionable and loses much of its plausibility if one distinguishes moral agency from moral considerability. While (self-)consciousness is clearly a requirement for moral agency and, hence, moral responsibility, there may well be beings who are not moral agents but whose interests must be considered in moral decisions.[3] This premise might be shored up with an additional argument that only moral agents can be morally considerable because only moral agents can participate in a recip-

---

[3] See James Rachels, "Do Animals Have a Right to Liberty?" and Tom Regan, "Do Animals Have a Right to Life?" in *Animal Rights and Human Obligations*, ed. Tom Regan and Peter Singer (Englewood Cliffs, N.J.: Prentice-Hall, 1975).

rocal moral community.[4] But this additional argument is plagued by questions of what, exactly, is the relevant sense of community involved. Are human beings who have lost the capacity for consciousness members of the moral community? If so, on what grounds, as they seem incapable of reciprocity or moral agency. Indeed, this whole line of reasoning is undercut by the widespread belief that moral agents can have obligations to humans who, for whatever reason, have lost the capacity for self-consciousness. Qualifications designed to circumvent such objections appear unsuccessful.[5] Consequently, defenders of this argument must deny the very widely held intuition that there do exist obligations toward permanently impaired, unconscious human beings. Most philosophers would find it more plausible to reject the claim that only conscious beings can be morally considerable, and, along with it, the implicit premise that only conscious beings can be moral beings.

Nor is it clear that only moral beings can have intrinsic value. Many aestheticians would argue that objects of art, for example, have intrinsic value but that they are neither moral agents nor morally considerable.[6] That is, moral agents may have no moral obligations *to* such works of art (though they may have obligations *regarding* them), and yet these works could have intrinsic value for purely aesthetic reasons. Thus, every premise of the proposed argument is questionable and likely to be rejected by some who hold that nonhuman species have intrinsic value.

Given the tenuousness of all the premises of the (self-)consciousness argument, one might wonder why it has been persuasive at all, much less considered a central argument for anthropocentrism. The explanation for its persistence lies in an ambiguity in the definition of intrinsic value, which was defined

[4] John Passmore, *Man's Responsibility for Nature* (New York: Charles Scribner's Sons, 1974), pp. 116-117.

[5] Tom Regan, "An Examination and Defense of One Argument Concerning Animal Rights," *Inquiry* 22 (1979): 189-219.

[6] See Stanley I. Benn, "Personal Freedom and Environmental Ethics: The Moral Inequality of Species," in *Equality and Freedom, International and Comparative Jurisprudence*, vol. 2, ed. Gray Dorsey (Dobbs Ferry, N.Y.: Oceana Publications, 1977).

in Chapter 1 as value an object has in its own right, independent of its value to any other object. But suppose a seemingly almost identical definition were put forward: intrinsic value is the value an object has in its own right, independent of any other object. On this definition one could argue that there is no valuing without a valuer and if only human beings (conscious beings, self-conscious beings) are valuers, nothing can have value in the absence of human beings (conscious beings, self-conscious beings). Therefore, no nonhuman object has value independent of humans and only they can have intrinsic value.

Some argument like this undoubtedly explains why anthropocentrists are attracted to the argument from consciousness.

But one could admit that all value is dependent on valuers in this sense—the sense that there is no value without a valuer—without admitting that the value of nonvaluing objects depends on the *values* of the valuer. That is, it requires the *act of valuing* by the valuer, but it does not depend upon the actual *values held* by the valuer. An object has instrumental value if its value depends on its contribution to the values of some other object. According to the definitions introduced in Chapter 1, however, a valuer can discover value independent of herself in some object that is not itself a valuer because intrinsic value is defined as value independent of the *values* of the valuer. The value of the object results from qualities it has rather than from its usefulness in serving values external to it. This, surely, is the conception of intrinsic value that would be employed by most believers in the intrinsic value of nonhuman species. For their purposes there is no need to defend the stronger and more controversial sense of independence whereby nonhuman species would have value even if there never had been (self-)conscious valuers.[7] To assume that the intrinsic value of nonhuman species must be regarded as wholly in-

---

[7] Given this clarification, it might be thought preferable to avoid the term "intrinsic" to describe the value attributed to nonhuman elements of nature, adopting instead the term "inherent." I choose, however, not to deviate from now deeply engrained usage, as too many environmentalists and critics have referred to and addressed the thesis that nonhuman elements of nature have "intrinsic" value. For a useful discussion of these points, see Robin W. Attfield, *The Ethics of Environmental Concern* (New York: Columbia University Press, 1983).

dependent of human valuers may prejudice the case against non-anthropocentrists.[8]

For all of these reasons the argument for anthropocentrism based on (self-)consciousness is inconclusive. All of its premises exploit contentious assumptions and definitions. A convincing argument deriving from this source would have to be based upon neutral premises, and such an argument appears unlikely to be valid.

7.3 *The Scriptural Argument*

Perhaps the most influential argument, historically speaking, for anthropocentrism is scriptural. It is written:

> And God said, Let us make man in our image, after our likeness; then let them have dominion over the fish of the sea and over the fowl of the air, and over the cattle, and over all the earth, and over every creeping thing that creepeth upon the earth.
>
> So God created man in his *own* image, in the image of God created he him; male and female created he them.
>
> And God blessed them, and God said unto them, Be fruitful and multiply, and replenish the earth, and subdue it; and have dominion over the fish of the sea, and over the fowl of the air and over every thing that moveth upon the earth.[9]

This is, of course, the thesis of human dominion over nature, which has been shown to be equivalent to anthropocentrism. The thesis is that all nonhuman creatures were created by God for human purposes. It has been taken to follow that only humans (in nature) have intrinsic value.

That the thesis of dominion has been enormously influential is indisputable. It was incorporated by René Descartes and Francis

---

[8] Paul Taylor, however, appears to defend this stronger version, which he refers to as "inherent worth." Personal communication, December 9, 1985. For a careful account of his concepts, see "Are Humans Superior to Animals and Plants?" *Environmental Ethics* 6 (1984): 149-160. See Chapter 1, note 7, above.

[9] Gen. 1:26-29, King James version.

Bacon as the heart of the modern scientific approach to the world. For example, Bacon described the goal of the scientific quest as allowing "the human race to recover that right over nature which belongs to it by divine bequest."[10]

Modern, secular science, the world view on which it rests, and the overwhelming modern faith in technology all depend on this foundational idea. However, it is worth noting, with John Passmore, that the idea of human dominion over nature did not develop into an aggressive, manipulative attitude until it was merged in the Renaissance with the Pelagian idea that the human race can, through art and knowledge, regain the dominion offered them at creation.[11] This hubristic attitude would have been rejected as heretical in the Middle Ages, along with the human attempt to perfect God's creation through science and technology, although the idea of human dominion was unquestioned at that time.

The thesis of dominion suffers from an ambiguity analogous to that which forced the distinction between strong and weak anthropocentrism above. It is possible to assert that nature's creatures are created for the good of man, without assuming a human right either to use them simply for the satisfaction of demand values or to extinguish them without moral guilt. One need not explain away Genesis I in order to attribute a role of stewardship to human inhabitants of the earth. Rather, insofar as, according to numerous other biblical texts, God charged the human race with being good stewards over his creation, he decreed ideals and rules of good and reasonable treatment of the earth and its other inhabitants. As Wendell Berry has explained, the very conditional and rule-circumscribed appropriation of the Promised Land provides a more accurate picture of the biblical view of the human relationship to the land because, unlike the creation story, it indicates God's view after the human fall from grace.[12] Attributing

[10] Francis Bacon, quoted from *The New Organon*, in William Leiss, *The Domination of Nature* (New York: George Braziller, 1972), p. 50.
[11] Passmore, *Man's Responsibility*, pp. 17-18.
[12] Wendell Berry, "Standing by the Words: The Biblical Basis for Ecological Responsibility," Sage Chapel Convocation, Cornell University, Ithaca, N.Y., November 11, 1979.

to humans a right of dominion and use of nonhuman species does not remove all restrictions on proper use. Human dominion over the earth need not imply human destruction of it.

Examined philosophically, rather than historically, what support is there for the thesis of dominion? The support seems merely to be that of scriptural authority. That it is only an argument from authority is damaging in itself. Of course advocates of the scriptural argument would claim a special status for it because it is *scriptural*. But what is to be made of this claim of special status? I will examine two answers—one traditional and the other cultural.

Claims of special status for the scriptural account traditionally rely upon scripture being the word of God. In the present case this amounts to saying that the scriptural account of creation embodies God's word that it is permissible to treat other species as if they have no intrinsic value. But many environmentalists believe there is strong evidence that actions based on this hypothesis are wrong actions. The traditional defender of anthropomorphism can, in the face of this evidence, do one of two things. Either he can (a) refuse to countenance the evidence, claiming that God's permission is always definitive, or (b) address the evidence with counterevidence. Refusal to countenance the evidence puts the full weight of the traditional case on (i) the claim that the scripture is the word of God and (ii) the claim that God doesn't authorize wrong actions.

But if the environmentalists do have strong evidence that actions consistent with God's edicts are wrong, they are of course committed to denying either (i) or (ii). So the debate between the environmentalist and the traditionalist will resurface as a debate about (i), (ii), or both. At that point, the traditionalist must, to defend his position, defuse the environmentalist's evidence—that is, he must opt for (b). But if he is eventually forced to provide counterevidence in order to support (i) or (ii), the biblical references to God's word have become otiose. The debate is, ultimately, about whether it is wrong to act as if nature has no intrinsic value. The argument is only made more complex by introducing God and God's decrees into the picture. The moral of the story is that either scripture can be mistaken or it cannot. If it can, then it doesn't help. If it cannot, then all disputants will claim to have it on their side. So one may as well deal directly with the evidence for specific, controversial claims.

I turn now to what I have called the cultural defense of the special status of the scriptural account. According to this view, the biblical account has special status as a central text in our culture: it expresses the essence of the Judaeo-Christian attitude toward nature and nonhuman species. To reject it would constitute a rejection of one of the most essential building blocks of the western world view.

Environmentalists may well accept this cultural claim—indeed, it amounts to little more than a paraphrase of Lynn White, Jr.'s claim that the Judaeo-Christian tradition, especially as embodied in the biblical creation story, is responsible for the growth of technology and the consequent destruction of large elements of nature.[13] But they will be quick to mention the wide array of current or predicted environmental crises and catastrophes as evidence that a culture deeply imbued with anthropocentrism mistreats nature. Projections that as many as one-fourth of all species now living may be extinct by the century's end are adequate evidence, they would claim, that any culture embracing anthropocentrism does so at terrible risk to nonhuman species. Far from providing support for anthropocentrism, this cultural argument would be taken by environmentalists to emphasize its evils. It may well be that the biblical account expresses a central idea of Western culture. But this idea may be one we would do well to reject.

7.4 The Evolutionary Argument

Theological justifications for ethical views have, of course, lost popularity in modern culture. W. H. Murdy has suggested a "modern" form of anthropocentrism that he believes to have been established by Charles Darwin's theory of evolution

[13] Lynn White, Jr., "The Historical Roots of Our Ecologic Crisis," in *Western Man and Environmental Ethics*, ed. Ian Barbour (Reading, Mass.: Addison-Wesley Publishing Company, 1973), pp. 18-30.

INTRINSIC VALUE

through natural selection. Murdy claims that Darwin dispatched the idea that nature exists to serve man and, simultaneously, established that "it is proper for men to be anthropocentric and for spiders to be arachnocentric. This goes for all other living species."[14] He goes on to quote approvingly G. G. Simpson's claim that anthropocentrism does not depend upon man's being the highest animal because, "even if he were the lowest animal, the anthropocentric point of view would still be manifestly the only one to adopt for consideration of his place in the scheme of things and when seeking a guide on which to base his actions and evaluations of them."[15]

The first of these claims, that other species do not exist for the good of man, follows, according to Murdy, from Darwin's conclusion that "natural selection cannot possibly produce any modification in a species for the good of another species."[16] This inference is not as obvious as Murdy suggests because it could be the case that natural selection within a species works for the survival of that species but that a transcendent being has arranged all the environmental factors affecting species so as to perpetuate those that are useful and necessary to human beings. But I only note this alternative in passing. The main concern of this chapter is with the second claim, that Darwin's theory of evolution through natural selection establishes anthropocentrism.

I turn now to an examination of Murdy's support for this claim. Murdy says:

> Species exist as ends in themselves. They do not exist for the exclusive benefit of any other species. The purpose of a species, in biological terms, is to survive to reproduce. Potter writes: "all successful living organisms behave purposefully in terms of their own or their species survival." Species that fail to do so become extinct. . . .
> To be anthropocentric is to affirm that mankind is to be

[14] W. H. Murdy, "Anthropocentrism: A Modern Version," *Science* 187 (1975): 1,168.
[15] Ibid. Quotation is from G. G. Simpson, *This View of Life* (New York: Harcourt, Brace, and World, Inc., 1947), p. 286.
[16] Murdy, "Anthropocentrism."

valued more highly than other things in nature—by man. By the same logic, spiders are to be valued more highly than other things in nature—by spiders.[17]

I will not quibble over Murdy's highly suspect implication that spiders are valuers. I take the point of this claim to be that the members of every species can, must, *and should* act so as to increase the survival chances of their species. While Murdy never spells out the argument in greater detail, I take it that these generalizations contain the main elements of his support for anthropocentrism.

Before attempting to develop Murdy's remarks into an argument, however, his position must be further clarified because of a qualification he adds later: "An anthropocentric attitude toward nature does not require that man be the source of all value, nor does it exclude a belief that things of nature have intrinsic value."[18] This surprising remark seems to belie my inclusion of Murdy as an anthropocentrist at all, by my definition. But it turns out that Murdy has a very unusual conception of intrinsic value. He explains, "I may affirm that every species has intrinsic value, but I will behave as though I value my own survival and that of my species more highly than the survival of other animals and plants."[19] Then he quotes an advocate of attributing intrinsic value to other species as saying that such a belief is necessary because, acting in his own self-interest, however enlightened, will not enable man to ensure "ecological survival."[20] Murdy's response to this claim is to accept it and to add: "Even this statement can be interpreted in terms of instrumental value, that is, man should acknowledge the intrinsic value of things: otherwise he will not have sufficient motivation for ecological survival, which I assume includes human survival individually and as a species."[21]

[17] Ibid. Quotation is from V. R. Potter, *Bioethics* (Englewood Cliffs, N.J.: Prentice-Hall, Inc., 1971).
[18] Murdy, "Anthropocentrism," p. 1,169.
[19] Ibid.
[20] Ibid.
[21] Ibid.

But this interpretation wholly subverts the claim that other species have intrinsic value. Murdy is suggesting that it is *true* that other species have intrinsic value but that humans are better off, practically, if they *believe* it and act as if this is so. This explains how he can consistently advocate that humans must (*and should?*) believe in the intrinsic value of other species at the same time that they act in their own interest—belief in the intrinsic value of other species is merely a ploy by which humans trick themselves into acting in ways that will be in humankind's long-term self-interest.

But what, exactly, is his argument for his central claim of anthropocentrism? The crucial premise derives from the theory of natural selection: members of all species *must* act in such a way as to protect their own individual lives and to perpetuate their own species. But how does one get from this premise to the conclusion that individuals are justified in always acting on human-oriented motives? Two possibilities arise. Either Murdy is suggesting that the matter is so wholly determined that individuals have no choice but to act as they do and so they cannot fairly be required to act otherwise—or alternatively he is suggesting that whatever scientific theories suggest individuals will do, they are morally encouraged or required to do (a naturalistic derivation of a moral claim from a fact).

These arguments appeal to different ethical principles and have different conclusions. Both start with the scientific premise that the theory of evolution through natural selection implies that individuals of all species will seek their own ends. The deterministic argument would build on this premise by appealing to the widely accepted ethical principle that "ought" implies "can." Employing this principle, self-serving behavior would be permitted because to prohibit it would be to require the impossible. The naturalistic argument appeals to the principle that people ought to do what scientific theories predict they will do. This principle would yield the stronger conclusion that individuals should seek their own interests, not just that they are permitted to do so.

The first option can be dismissed out of hand. If Murdy believed that all individual actions were determined inflexibly by

ANTHROPOCENTRISM

behavioral signals coded into our genes, there would be no point in writing an article clarifying the motives by which people act and there would be no point in advocating changes in human behavior as he does later in his essay. This ad hominem aside, the argument is unattractive because it cuts the ground out from under any moral argument. If humans have no free choice, the conclusion of Murdy's argument—that humans are morally permitted to act on purely human interests—would be meaningless. Moral pronouncements have meaning only in a context where a real choice is involved, and determinism precludes choice.

It is tempting to reject the second interpretation of Murdy's argument just as briefly, since it appears to exemplify the "naturalistic fallacy." It is generally thought that moral conclusions cannot be derived from purely factual assertions such as scientific theories. But perhaps the argument is worthy of a closer look. There may be some plausible moral principle that could be supplied as an added premise to get from the facts of natural selection to the claim that human beings are justified in acting on purely anthropocentric motives.

What might such a premise be? I have isolated two candidates:
(A) If members of nonhuman species act to perpetuate their lives and their species, then humans should do so as well (a sort of universalizability principle applied across species).
(B) Unless members of each species act to perpetuate their lives and their species, natural selection (the process by which each species adapts to its changing environment) would cease to operate, and it is a good thing for natural selection to continue.

One might, then, go from either (A) or (B) to the conclusion that human beings are morally obligated to act to perpetuate their lives and their species. In each case the addition of a nonfactual premise makes possible the inference to the moral conclusion.

Premise (A) gains any plausibility it has from the principle that moral rules should apply equally to all equal cases. It might here be reasoned that since all other species are engaged in a tooth-and-claw struggle, humans could not fairly be asked to put their species at a disadvantage by playing under a different set of rules.

147

This addition to Murdy's argument has the exegetical advantage of finding a role for his remarks about "arachnocentrism" and the behavior of members of other species that otherwise seem irrelevant.

But the universalizability principle applies only if there is reason to believe that the cases are equal; and here they are not. Members of other species, it seems safe to conclude, have no free will and cannot help how they act. Their behavior is determined by instinct or stimulus-response conditioning. If humans are determined as well, this form of the argument collapses into the deterministic argument rejected above. But if human beings have free will, then it follows that their behavior is not relevantly similar to that of members of other species. Human beings are responsible actors to whom moral principles and distinctions are relevant. If other species have intrinsic value, human beings have an obligation to consider that value in deciding what to do, whether or not it is in their self-interest to do so.

The alternative form of the argument, which exploits premise (B), is reminiscent of social Darwinist arguments, which are now given little credence. Premise B surely requires careful qualification, as it would rule out many widely accepted practices. We may believe, in very general terms, that, as (B) asserts, it is good for natural selection to continue to operate. But we make so many special exceptions to it—by, for example, wearing eyeglasses, correcting operable congenital defects in human children, and selectively breeding other species for our own good—that the principle could be used in an argument such as Murdy's only with careful qualification. This premise seems especially dubious in light of Murdy's admission that humans, failing to recognize (or pretend) that other species have intrinsic value, are in danger of extinguishing themselves. Indeed, it is ironic that the entire thrust of the later parts of Murdy's essay is to argue that humans should pay more attention to the long-term good of the species because present actions threaten the human race with extinction. Presumably, present actions are motivated by selfishness—the very motive Murdy is espousing.

The only way out of this puzzle is to distinguish (a) individual selfish motives, (b) motives based on protection of the human race in the long run, and (c) motives based on human ideals (perhaps including altruism toward other species). Once these distinctions are made, the present interpretation of Murdy's argument implies that the natural tendency toward (a) should be curbed in favor of (b), even if this requires that humans pretend to believe in (c). But where does this leave "natural" selection?

Once conscious evaluation of motives enters the picture, arguments that one ought to let nature take its course become confusing. Which motives are, in fact, anthropocentric? The individual, primal, and instinctual urges? The consumptive actions fueled by these are the basis of the environmental crisis from which Murdy is trying to free the human race. Those directed toward the rationally projectible long-term good of the species? This suggestion raises difficult questions. No other species consciously chooses to sacrifice individual consumptive needs for the long-run good of the species. Indeed, natural selection normally works insofar as each individual seeks to maximize his own chance of survival. Cases where individual organisms jeopardize their own survival to improve the chances of the group, as when army ants drown to make a bridge for other ants, are thought peculiar and require an unusual evolutionary explanation. An emphasis on "natural" individualistic motives defeats Murdy's conclusion that humans need to protect the species in the long run.

Perhaps Murdy's argument could be saved, however, by noting that human reason, with its ability to alter the world, has resulted in environmental destruction, but the same intellectual powers allow human beings to recognize oncoming disaster and to plan for the long-run protection of the species. That is, each species can be expected to make the best of the repertoire of skills and behavior available to it; humans have rational consciousness and they too should use it. But this argument still fails to provide any reason why individuals should choose to protect the human race in the long run at the expense of their own individual interests. It is true that humans can choose to act on ideals such as the protection of the race, but the requirement to limit individual competitive activity by the conscious institution of ideals does not follow from Darwinism. Additional philosophical arguments showing that these ideals take precedence over individual satisfactions must be adduced as well. But once the conscious institution of ideals is raised as a possibility, why not make those ideals altruistic toward other species? Murdy's suggestion that Darwinism supports survival of the human race as the central goal *and* that all altruistic ideals must be interpreted as instrumental to that goal turns out, then, to be groundless.

In this chapter several justifications for anthropocentrism have been discussed and rejected. Most of the other lines of reasoning that are purported to support this conclusion have been subjected to seemingly devastating criticism elsewhere.[22] Since the preemptive strikes of anthropocentrists against the thesis that other species have intrinsic value have failed, I will turn, in the next two chapters, to an examination of attempts to make sense of and justify the claim that other species have intrinsic value. Next I will discuss the highly popular strategy of interpreting the claim that other species have intrinsic value individualistically. Then I will examine nonindividualistic approaches.

[22] See Routley and Routley, "Against the Inevitability."

E I G H T

NONANTHROPOCENTRISM I:
INTRINSIC VALUE AND INDIVIDUALS

8.1 Direct and Indirect Analogies

A successful case for anthropocentrism would render useless attempts to establish the intrinsic value of nonhuman natural objects. But, as we have seen, the arguments for anthropocentrism have proved inconclusive. So the question now arises: Can such attempts be given a plausible theoretical understanding?

It will be useful first to classify the numerous possible positions by their answers to three questions:
(1) What is the *locus* of the intrinsic value ascribed?
(2) What does it *mean* to ascribe intrinsic value to an object?
(3) How are ascriptions of intrinsic value to nonhuman objects *justified*?

There are three reasonable answers to the question of locus that, in different combinations, yield seven possibilities. Intrinsic value can be attributed: (1) to individual members of nonhuman species; (2) to species; (3) to ecosystems; (4) to individuals and species; (5) to individuals and ecosystems; (6) to species and ecosystems; and (7) to individuals, species, and ecosystems. Although not all of these positions have been defended explicitly, it is useful to recognize that species preservationists can choose from a range of possible loci for intrinsic values supporting the protection of species.

Because there exists no single, accepted conception of intrinsic value, a variety of answers are available as well to question (2). The various accounts of intrinsic value share a general definitional schema: the intrinsic value of an object is that value it has

151

which is not dependent on its contribution to the value of another object. But they differ in what positive account they propose of the characteristics an object must have if it is to qualify as intrinsically valuable. Indeed, they differ as to whether or not they offer a positive account at all. I will discuss presently some of the positive answers offered by environmental philosophers. For the purpose of classifying theories, however, I will contrast approaches that include such a clarification with those that do not.

Many proposals that describe nonhumans as intrinsically valuable fail to offer any positive clarification of their central notion. Proponents of these proposals, rather, attempt to *justify* their positions, that is, offer an answer to question (3), bypassing the seemingly prior question of exactly what it is they are justifying. They argue that some nonhuman item has intrinsic value by, first pointing to some thing that is generally recognized as having intrinsic value and, second, arguing that there is no relevant difference between that thing and the proposed nonhuman item.

For example, such an argument might be: (1) Individual humans are attributed intrinsic value; (2) There is no difference between nonhuman and human individuals that is relevant to attributions of intrinsic value; therefore, (3) Nonhuman individuals must be attributed intrinsic value. In this way (which I will refer to as "denying the disanalogy"), the advocate of intrinsic value can avoid specifying the positive characteristics that entail intrinsic value. Also, by using human intrinsic value as the accepted reference point of the analogy, those who deny any disanalogies suggest a parallel with civil rights arguments based on the moral equality of all humans. Opponents of intrinsic value of nonhumans are labeled as "speciesists," "human chauvinists," etc.[1] They are accused of refusing to grant intrinsic value to other species arbitrarily, on no basis other than that they are different species. And since those who merely deny disanalogies give no positive

[1] See, for example, Peter Singer, "Not for Humans Only: The Place of Nonhumans in Environmental Issues" and Richard Routley and Val Routley, "Against the Inevitability of Human Chauvinism," both in *Ethics and Problems of the 21st Century*, ed. K. E. Goodpaster and K. M. Sayer (Notre Dame, Ind.: University of Notre Dame Press, 1979).

definition of intrinsic value, there is no criterion or characteristic that one can challenge if one wishes to dispute the proposed attribution of intrinsic value. "Intrinsic value" equals "whatever it is that humans have that justifies one in saying they have noninstrumental value." The method of denying the disanalogy is popular among nonanthropocentrists. It may seem as if this method gains for its users considerable advantages in argument. But these advantages are the ill-gotten gains of a retreat into vagueness.

Some nonanthropocentrists have been more bold, however, offering positive clarifications of the attributions they make before justifying them. This approach situates claims that nonhumans have intrinsic value in a general value theory including some positive interpretation of what intrinsic value is and what characteristics impart it to any object. Justification can then proceed by arguing that the proposed possessor of intrinsic value has the characteristics indicated as the definitive mark of that type of value. According to this approach, direct analogical reasoning might be helpful—it could be argued that humans, for example, have the relevant characteristics and that the proposed possessor of intrinsic value is similar to humans in the pertinent respects.

But the use of positive analogy is not essential. References to humans and human characteristics serve only to illustrate the point and could be eliminated in favor of a straightforward argument asserting a defining characteristic for intrinsic value and a further argument that the proposed object has this relevant characteristic. In denials of disanalogy, analogical references are not similarly eliminable.

It is useful, then, to distinguish arguments for nonanthropocentrism as indirect (denials of disanalogy) or as direct (whether analogical or not). Positions employing only the indirect method give no positive clarification for the central notion of intrinsic value; they justify nonanthropocentrism without clarifying it. The method of direct argument, whether analogical or not, must accept and defend some positive conception of intrinsic value.

Using this dichotomy in conjunction with the seven possible loci of intrinsic value listed in the answer to question (1), it is possible to chart fourteen categories of theories of intrinsic value (see

153

chart). Indeed, there are many more possibilities than that, because several different axiological theories might be used in attempts to clarify and justify any of the seven options. Not all possible positions have been defended in the literature, and my treatment will not attempt an exhaustive discussion of the possibilities. A significant portion of the possible positions seem initially unattractive, and some economy will result because theories combining two or three loci of value will often suffer all of the problems attributed to the independent loci they comprehend.

	A. Indirect Method (essentially analogical)	B. Direct Method (analogical or not)
1. Individual members of nonhuman species	Individual humans: Individual nonhumans (Taylor[2])	Individual nonhumans have C (where C refers to some characteristic constitutive of intrinsic value) (Singer[3])
2. Species	Human species: Nonhuman species	Nonhuman species have C (Callicott[4])
3. Ecosystems	Human communities: Ecosystems	Ecosystems have C (Callicott[5])
4. Individuals and species	Individual humans: Individual nonhumans	Nonhuman individuals have C
	Human species: Nonhuman species	Nonhuman species have C (Callicott[6])

[2] Paul W. Taylor, "The Ethics of Respect for Nature," *Environmental Ethics* 3 (1981): 197-218. Taylor believes nonhuman individuals, nonhuman species, and ecosystems have intrinsic value. Since he bases the latter two on individual value, however, it seems fair to refer to him as a representative of A.1 as well as A.7.

[3] Singer, "Not for Humans Only."

[4] J. Baird Callicott, "On the Intrinsic Value of Nonhuman Species," in *The Preservation of Species*, ed. Bryan G. Norton (Princeton, N.J.: Princeton University Press, 1986).

[5] J. Baird Callicott, "Animal Liberation: A Triangular Affair," *Environmental Ethics* 2 (1980): 311-338.

[6] Callicott, "Intrinsic Value." I have placed Callicott's biosentimentalism here

5. Individuals and eco-systems	Individual humans: Individual nonhumans	Nonhuman individuals have C
	Human communities: Ecosystems	Ecosystems have C
6. Species and ecosystems	Human species: Nonhuman species	Nonhuman species have C
	Human communities: Ecosystems	Ecosystems have C
7. Individuals, species, and ecosystems	Individual humans: Individual nonhumans	Nonhuman individuals have C
	Human species: Nonhuman species	Nonhuman species have C
	Human communities: Ecosystems (Taylor[7])	Ecosystems have C (Regan[8])

In the remainder of this chapter I will discuss several attempts to clarify and justify attributions of intrinsic value to nonhuman species deriving from the analogy to intrinsic value as found in human and nonhuman individuals. These approaches attempt to transfer the intrinsic value initially located in individual human specimens to species, but they treat individuals as the locus of all intrinsic value. In Chapter 9, I will examine attempts to attribute intrinsic value directly to collectives such as species and eco-systems.

Before assessing these positions, it will be useful to state my evaluative strategy. I will insist that any nonanthropocentric theory must fulfill two conditions: first, any intrinsic value attrib-

as well as in category B.2 because it is unclear whether he believes that nonhuman individuals have intrinsic value. His style of argument certainly suggests this, but he never says so explicitly, and in "Animal Liberation," an earlier essay, he seems to deny such attributions (pp. 324-325).

[7] Taylor, "Ethics of Respect."

[8] Tom Regan, "The Nature and Possibility of an Environmental Ethic," *Environmental Ethics* 3 (1981): 19-34. While Regan specifies no positive characteristic constitutive of intrinsic value, he clearly states that such a specification should be the objective of environmental ethicists.

uted, either to objects serving as the basis for the analogy or to objects by analogy, must be capable of explanation and justification under some plausible theory of intrinsic value. This requirement allows reference to intrinsic value that is not fully explained and justified theoretically, but it rules out such claims if there are persuasive reasons to believe they could *never* be so explained or justified. This feasibility condition is intended to ensure that the claims be more than subjective appeals to moral feelings. Second, the type and range of intrinsic value attributed must be sufficient to justify characteristic claims by species preservationists. This adequacy condition requires that the attributions of intrinsic value proposed are adequate to the task at hand.

While this second condition relativizes any criticisms and conclusions to a specific purpose, it should be recognized that the purpose of preserving species is a very central one in the broader picture of environmental protectionism. Concern for biological diversity stands, in a sense, as the most central value of environmentalism because other environmental goals such as resource protection, pollution abatement, and so forth all depend upon the continued functioning of complex ecosystems.

8.2 *Intrinsic Value and Individual Interests*

The least controversial case of intrinsic value is the value found in an individual human life. The most straightforward analogy available to nonanthropocentrists, therefore, takes human individuals as its reference point and reasons that nonhuman individuals are sufficiently similar to human ones to justify similar attributions. A long and persuasive tradition affirms that individual human beings are not to be treated purely as means and that they must, in ethical decisions, be considered as ends-in-themselves. Concern for individual nonhumans, taken on analogy to concern for human individuals as ends-in-themselves, amounts to concern for their individual welfare, for their rights or interests. Differing ethical traditions place differing weight on rights and interests, but for present purposes they can be used interchangeably: they refer to the basic moral currency attributable to individuals and deriving from the status of those individuals as ends-in-themselves.[9]

Environmentalists often speak as if claims that nature has intrinsic value mean no more or less than that nonhuman creatures have rights. Two much-quoted and much-discussed passages from Aldo Leopold's classic appeal for a new "land ethic" illustrate this tendency:

> The Golden Rule tries to integrate the individual to society; democracy to integrate social organization to the individual.

> There is as yet no ethic dealing with man's relation to land and to the animals and plants which grow upon it. Land . . . is still property. The land-relation is still strictly economic, entailing privileges but no obligations.

> The extension of ethics to this third element in human environment is, if I read the evidence correctly—an evolutionary possibility and an ecological necessity.[10]

and

> A land ethic of course cannot prevent the alteration, management and use of these "resources," but it does affirm their right to continued existence, and, at least in spirit, their continued existence in a natural state.[11]

Leopold's eloquent argument moves smoothly from concern for human individuals to attributions of rights to a wide range of nonhuman objects. But two distinct propositions could be isolated from these passages.

(1) Humans have an obligation to respect nature and, in particular, to protect the existence of other species.

---

[9] Deontologists favor rights as the currency of individualist ethics; utilitarians favor interests. Taylor prefers the term "welfare," but this seems mainly motivated by his desire to emphasize that an individual can have a welfare without "taking an interest" in it. Leaving this point aside, his conception of welfare seems very similar to the usual concept of interest ("Ethics of Respect," pp. 199-200).

[10] Aldo Leopold, *A Sand County Almanac* (London: Oxford University Press, 1949), p. 203.

[11] Ibid., p. 204.

(2) Nature and, in particular, other species have a right to continued existence.

Environmentalists often speak of (1) and (2) as if they were interchangeable. But one can consistently accept (1) and reject (2).[12] If
any object has intrinsic value, it follows that one has an obligation to protect it (if possible). It does not follow that the object
has rights in any technical sense of that term. The logical geog-
raphy, then, is as follows: a belief that nature and other species
have intrinsic value entails (1) but not necessarily (2). A belief in
the rights of nature and other species, that is, (2), entails (1) but
not vice versa.

It will be convenient first to examine the cogency of arguments
for (2) understood not only as ascribing rights in the usual sense
but in the broader sense of attributing morally relevant interests
(moral considerability) to nonhuman species. If such a claim can
be justified, it would settle the question of intrinsic value positively. Further, since the concept of rights appears at least on the
surface to be relatively clear in some contexts, a successful defense of rights for nonhuman species would go some way toward
clarifying the concept of intrinsic value. As Callicott says, appeals
to rights can be seen as the "preferred locution" of environmentalists who believe that other species have intrinsic value.[13] Rights
are the strongest moral currency—in Ronald Dworkin's terms,
they are moral "trumps," overriding other sorts of concerns such
as convenience.[14] Rights can be overridden only by other rights of
higher priority. So it is not surprising that liberation movements,
anxious to expose engrained prejudices and to redress longstanding grievances, express their viewpoints in the strongest available
language. In this sense appeals to rights may be calculated over-
statements. They may have purely forensic force.

But forensic appeals are not sufficient for the present purpose.
The goal here is to understand if and why it is necessary for humans to protect other species, to examine the various reasons

[12] Callicott, "Intrinsic Value."
[13] Ibid.
[14] Ronald Dworkin, *Taking Rights Seriously* (Cambridge, Mass: Harvard University Press, 1977).

given to preserve species. Liberation movements are convinced of the correctness of their moral views and are looking only for stirring language to convince others. The task here, however, is to justify claims in a philosophical manner, and this process requires that appeals to concepts such as rights, interests, and intrinsic value be justified within a plausible theory of value.[15] Otherwise, appeals to these concepts function as no more than subjective invocations of moral feelings.

I will examine two attempts to respond to the challenge implicit in these remarks. Paul Taylor, using the indirect method just described, denies any disanalogy between nonhuman and human welfare.[16] He argues that individual, living nonhumans have a welfare of their own and that failures to take their interests into account are based on an irrational and arbitrary bias in favor of humanity.[17] Peter Singer, using a direct approach, has argued that consciousness, or sentience, is sufficient to entail intrinsic value and moral consideration to its possessor, thereby generating a positive analogy from human interests in feeling pleasure and avoiding pain to similar interests in other sentient beings.[18] Because Singer's approach involves clear and concrete assertions about intrinsic value, I will discuss his argument first. Taylor's approach will be discussed in the next section.

Singer argues that "consciousness, or the capacity for subjective experience, is both a necessary and a sufficient condition for

---

[15] Ruth Macklin, "Moral Concerns and Appeals to Rights and Duties," *Hastings Center Report* 6 (1976): 31.

[16] I have classified Taylor as using the indirect method based upon a reading of the essay "The Ethics of Respect for Nature." It is true that he identifies the characteristic of "having a good of one's own as central to the justification of attributing inherent worth to a being." But he denies that "there is a logically necessary connection between the concept of a being having a good of its own and the concept of inherent worth" (p. 204). As I read the essay, attributing inherent worth to nonhumans in addition to attributing a good of their own to them relies on the "symmetrical" relationship with human ethics (p. 206) and the denial that humans are "superior" to animals and plants (pp. 207-208). I understand that in Taylor's book *Respect for Nature* (Princeton, N.J.; Princeton University Press, 1986), the reasons for attributing inherent worth are spelled out more positively and in more detail. The full text of Taylor's book was not available to me at this writing, however.

[17] Taylor, "Ethics of Respect," pp. 199, 216.

[18] Singer, "Not for Humans Only," pp. 194–195.

having an interest."[19] Without claiming that all conscious beings have the same interests, he argues that the principle that all interests must be given equal consideration (the basis of equality among humans) must be extended to all conscious beings.[20] Only in this way, he believes, can unjustified discrimination be avoided. However different individual conscious beings and their interests may be, equal consideration is due them, and like interests should be treated alike. The only justification for different treatment is a difference in interests.[21]

Without doubt, this approach fulfills the feasibility condition set out in the last section: Singer's attribution of intrinsic value to all conscious beings can be explained and justified within a plausible theory of intrinsic value. It shows, also, some initial promise for extending moral concern to important elements of nature. If humans hunt other species for sport, if they destroy large areas of habitat, if they modify the face of the earth with no concern for the lives of members of other species, they surely can be accused of ignoring the interests of nonhumans. Reformulating human policies to show sensitivity toward these concerns would go some way toward eliminating human destruction of other species.

But Singer's approach falls short of providing a satisfactory theoretical underpinning for a comprehensive policy of species protection for at least two reasons. First, even if one accepts Singer's attribution of consciousness, interests, and, consequently, intrinsic value to many animal species, this sort of reasoning cannot provide a basis for protecting plant or less complex animal species. Concern based on the positive analogy deriving from human sentience can extend only to species sharing that characteristic. Since species preservationists want to save species ranged throughout the phylogenetic scale, many of which lack consciousness, Singer's approach cannot provide comprehensive support for preservationist policies based upon the interests of threatened individuals.

[19] Ibid., p. 194.
[20] Ibid. Also see Lily-Marlene Russow, "Why Do Species Matter?" *Environmental Ethics* 3 (1981): 101-112.
[21] Singer, "Not for Humans Only," p. 195.

Second, since Singer's approach generates concern for nonhuman species through an analogy between individual human characteristics and individual nonhuman characteristics, he cannot generate a special concern for endangered species.[22] Practices that favor members of endangered species are integral to the program of species preservationists, however, as can be seen by examining a few examples.

Species preservationists have, on occasion, advocated the removal or even the destruction of large herbivores such as goats and burros from islands because the animals are so destructive of indigenous vegetation. Since plants cannot have interests, concern for interests alone would dictate protecting the goats and burros at the expense of even highly endangered plant species.

Species preservationists also sometimes reluctantly advocate culling of herds of endangered species, since protected populations of endangered species, confined within limited habitats, sometimes experience local population booms that threaten to overconsume their food supply. The result may be destruction of their food source and danger of massive die-offs, increasing the likelihood of extinction. Singer's approach could never support a policy of culling herds, as individual interests are paramount on any system of ethical obligations to nonhumans derived in analogy to obligations to human individuals.

When the blue whale became exceedingly rare, species preservationists supported laws protecting them, even though they were well aware that these laws would increase pressure on more abundant whale species such as the sperm whale. They could not justify this support on the grounds that blue whales, taken individually, have weightier interests, as members of the two species have essentially equivalent levels of consciousness and, consequently, their interests should be accorded essentially equal treatment.

Singer recognizes the limitations inherent in his individualistic approach, as he asks: "On what basis, then, other than the indi-

---

[22] See Elliott Sober, "Philosophical Problems for Environmentalism," in *The Preservation of Species*.

rect benefits to humans, can we justifiably give preference to the preserving of animals of endangered species rather than animals of species that are not in any danger?"[23] He examines and rejects several answers as indefensible or incomplete and says, "To this question I can find no satisfactory answer. . . . I conclude, then, that unless or until better grounds are advanced, the only reasons for being more concerned about interests of animals from endangered species than about other animals are those which relate the preservation of species to benefits for humans and other animals."[24] There is no reason to prefer fifty whooping cranes to fifty sandhill cranes, except for human reasons. This is tantamount to admitting that species protection cannot be fully supported by appeals to interests of nonhuman individuals—some important species would not be protected on Singer's approach. Nor can it provide the central theory justifying species preservation.

An emphasis on nonhuman individual interests and rights derived by positive analogy from human individual interests and rights fulfills the feasibility condition but not the adequacy condition. Singer's theory, although difficult to defend conclusively, provides a coherent account of intrinsic value deriving from individual consciousness of pleasure and pain. All conscious beings have intrinsic value in the sense that their interests in experiencing pleasure and avoiding pain must be taken into account in ethical decision making. Species preservationists may take some comfort in this conclusion; but purely individualistic approaches, especially ones anchored in the analogy to human consciousness, cannot support the full range of programs necessary for a comprehensive policy of species preservation.

8.3 Intrinsic Value and Nonhuman Welfare

Perhaps in recognition of these problems, Paul Taylor has developed a much more comprehensive system of attributing intrinsic value to nonhumans. He ascribes intrinsic value to all living

[23] Singer, "Not for Humans Only," p. 203.
[24] Ibid.

things as individuals, to species, and to natural systems.[25] But, like Singer, he bases his value system on concern for individuals, arguing that "finally it is the good (well-being, welfare) of individual organisms, considered as entities having inherent worth, that determines our moral relations with the Earth's wild communities of life."[26] A theory such as this seems, initially, to hold more promise to fulfill the adequacy condition; it requires, however, a correspondingly more comprehensive conception of intrinsic value. This more comprehensive conception rests, like Singer's, on an analogy derived from the human case, but Taylor uses the analogy indirectly—he does not identify a central characteristic such as sentience as determinative of intrinsic value. He argues only that the case for human superiority is groundless; it is "at bottom nothing more than an irrational bias in our own favor."[27]

Because he wishes to apply it across the entire spectrum of species, Taylor's conception of intrinsic value cannot rest on consciousness, sentience, or the experience of pleasure and pain. He adopts a broader conception of an individual's welfare or good of its own, a conception not entailing that the being has interests or takes an interest in what affects its life.[28] He argues that intrinsic value ("inherent worth") should be attributed to the realization of the good of any individual that exists as a "teleological center of life."[29]

I say that Taylor uses the indirect method of denying the disanalogy because, while he believes that being a teleological center of life is shared by all living things and that the recognition of this characteristic is an important step toward attributing intrinsic value, his arguments for taking this further step rely heavily upon denying disanalogies to human beings. Singer rests his assertion that we should promote feelings of pleasure and minimize feelings of pain (wherever they occur) on a general, utilitarian theory

---

[25] Taylor, "Ethics of Respect," p. 198.
[26] Ibid.
[27] Ibid., p. 215.
[28] Ibid., p. 199. See note 9, above.
[29] Ibid., pp. 201, 211.

of value. Taylor argues only indirectly that there is no reason to withhold attributions of intrinsic worth from nonhumans, provided they are affirmed for humans, because there are no unbiased reasons to favor humans in these regards.

Taylor is thereby able to avoid furnishing a positive conception of intrinsic value or committing himself to any set of necessary and sufficient conditions for its possession. But the costs attendant upon this approach are high. No positive reason for attributing intrinsic value to nonhumans can be offered, and his entire case, aside from negative attacks on arguments for human superiority, must gain its plausibility from unspecified similarities between humans and members of other species. Given the incredible range and variety of life forms, the perceived similarities available to anchor Taylor's theory appear minimal.

When Taylor's system is viewed from this perspective, a legitimate reaction would be that the reader is being asked to accept a very great deal based upon very little evidence. Besides leaving the characteristics constitutive of individual intrinsic value largely unspecified, Taylor has also apparently assumed that these characteristics, whatever they are, are the same characteristics that confer intrinsic value on collections of individuals such as species and ecosystems. That is, the value of species and ecosystems is derivative upon the value of individuals composing them.[30] But it became clear in the discussion of Singer that purely individual values cannot generate a special value for endangered species.

If species have intrinsic value because of the intrinsic value of their members on analogy to the intrinsic value of human individuals, then severe limitations must be placed on the management techniques available to preservationists. For example, herd culling could be justified according to this approach only if we were equally well prepared to exterminate human individuals when

---

[30] Ibid., p. 199. But the inference that species have value because their individual members do has also been challenged. See, for example, Jan Narveson, "Utilitarianism and New Generations," *Mind* 76 (1967): 72; and Joel Feinberg, "The Rights of Animals and Unborn Generations," in *Philosophy and Environmental Crisis*, ed. William Blackstone (Athens, Ga.: University of Georgia Press, 1974), p. 66.

populations exceed their carrying capacity. Since this policy would be decried as absolutely immoral in the case of humans, it could not be applied to other species, given Taylor's analogy. Perhaps this point is not damaging, as Taylor might convince the species preservationists that this practice is, indeed, immoral and should be stopped.

But what management techniques would remain? To the extent that the value of species is, according to Taylor's analogy, derived from the value of individuals, policies that give special treatment to individuals of an endangered species would constitute unjustified discrimination against individuals of other, nonendangered species. If the intrinsic value of a species is the aggregated intrinsic value of its members, as Taylor claims, then the value of the species does not exceed the collective value of its members. How, then, can species preservationists use the intrinsic value of nonhuman species (so justified and understood) to support policies that admittedly discriminate in favor of some individuals on grounds derived from characteristics of their species? It appears, then, that species preservationists cannot attribute intrinsic value to nonhuman species by aggregating interests of individuals. To serve the case of species preservationism, the intrinsic value of species must be derived from some source indepdendent of the intrinsic value possessed by their members.

Even so, the intrinsic value attributed to specimens will conflict with, and limit the means to secure, the protection of species. Treating individual creatures with special care simply because they are specimens of endangered species and withholding similar treatment from specimens of nonendangered species discriminates in favor of and against individuals as a result of characteristics of their group. When one justifies differential treatment of individuals because of the status of their species, one treats individuals as means to the preservation of species and denies that they are ends-in-themselves.

This is, surely, a damaging conclusion for anyone who hopes, as Taylor does, to base protectionist policies on claims of intrinsic value of individuals. It is not just that intrinsic value must be attributed to each species independently of the value of its members; the value of the members will often be in conflict with the value of the species. Concern for individuals flows against the tide of concern for species. Taylor's approach, then, suffers two serious problems. First, it rests on questionable support: no clear and positive basis for its negative analogical claims is advanced. Second, and even more damaging, the approach cannot justify centrally important policies of species preservation.

8.4 The Bankruptcy of Individualism in Species Preservationism

Theories attributing intrinsic value to nonhumans have tended to involve, in one way or another, the attribution of intrinsic value to nonhuman individuals. Reasons for this tendency are not far to seek. The intrinsic value of human individuals stands as the most clear and least debatable case of intrinsic value. It is not surprising, therefore, that this case has been chosen as the model for new and more controversial attributions. It is also true that human compassion for other species is at least originally directed toward individuals. Individualistic theories can then build on this compassion for members of other species.

Further, modern approaches to ethical theory are predominately individualistic. Utilitarianism and deontology are individualistic in the sense that both start with interests or rights of individuals as their most basic currency. Both utilitarians and deontologists can understand the value of an individual human as a being whose welfare is worth promoting, other things being equal. Appeals to rights or interests of members of nonhuman species retain the form of such ethical systems while merely broadening the reference class. If one thinks of animal liberationists and species preservationists as ethical pioneers, it is a great advantage to them to be able to transfer a relatively clear operating system of ethical concepts and rules into their new area of concern.

Finally, as was noted in section 8.2, appeals to rights and morally relevant interests are the strongest moral currency. If species protectionists could explain their goals in terms of these concepts, they would simultaneously provide a clarification of attributions of intrinsic value to nonhuman species and gain plausibility for their case by emphasizing its parallels with historically successful liberation movements.

But while these reasons conspire to make the individualist approach attractive to defenders of intrinsic value of nonhuman species, attempts to develop such accounts in detail have proved disappointing. It is very difficult to move from concern for individuals to concern for species and, worse, to the extent that individualistic approaches invest specimens with justifiable claims to equal consideration of their interests, these claims militate against discriminatory practices usually assumed to be necessary in a comprehensive policy of species preservation.

I hope my discussion of representative approaches, though admittedly not exhaustive, accentuates the inherent difficulties in deriving intrinsic value of species from intrinsic value of nonhuman individuals, especially when that value is modeled on the intrinsic value of human individuals. Concern for individuals originates in human compassion regarding pain and death felt by other organisms. This compassion is not evenly distributed throughout the phylogenetic scale, so it must somehow be extended as well as justified if it is to provide a comprehensive moral basis for species preservation. Extensions are only as plausible as the analogy to human rights, interests, and welfare on which they are based, and the broader one construes the analogy, the less plausible it becomes to argue that nonhuman individuals share with human ones some central characteristic affording them equal or similar moral status.

The case can be made even more powerfully. Concern for welfare of individuals seems, in general, to be the wrong direction to look if the goal is to preserve species. As Sagoff has pointed out, the best means to protect individual members of nonhuman species from pain and death is to remove them from the wild and place them in zoos and botanical gardens.[31] But this weakens the

[31] Mark Sagoff, private conversation. Also see Sagoff, "On Preserving The Natural Environment," *Yale Law Journal* 84 (1974): 221–222.

forces of evolution that create diversity and promote adaptation, that strengthen species and increase their chances of surviving.[32]

Every sufficiently lucrative development project that infringes on the habitat of a species could be accompanied by a plan to ensure that no individual interests of any member of an endangered species would suffer harm as a result of the project. Members of the species in question could be painlessly sterilized and then placed in a carefully controlled habitat to live out their natural lives in luxury. Individualistic concerns for rights or interests could, I am convinced, be handled in an ad hoc manner, given enough money and effort. But this does not protect the species, an ongoing entity that preservationists hope will continue through indefinite time, or at least for many generations.

These difficulties belie the initial attractiveness of the individualist approach. Referring to the chart on pages 154-155, it would appear that all theories that generate the intrinsic value of species from the intrinsic value placed on individuals can be eliminated from serious consideration as a support for species preservation. This conclusion applies to most theories in categories A.1, B.1, A.4, B.4, A.5, A.7, and B.7.[33] I will turn in the next chapter to attempts to explain and justify claims that nonhuman species have value noninstrumental to human values but that are not derived from the value of their individual members.

---

[32] Callicott, "Animal Liberation," pp. 332-336.

[33] I say *most* theories because it is possible to defend a theory that places intrinsic value on nonhuman species and/or ecosystems *in addition to* individuals but does not base the value of species and ecosystems on the value of individuals. Problems with such an approach are legion, including the lack of a consensually accepted basis for nonindividual intrinsic value and the introduction of competing values, but such a theory is conceptually possible. At any rate, I have shown that species preservationists gain no advantage, indeed damage their cause in some respects, by establishing that nonhuman individuals have intrinsic value.

N I N E

NONANTHROPOCENTRISM II: SPECIES
AND ECOSYSTEMS

9.1. *Intrinsic Value and Collectives*

My arguments have cast doubt on the usefulness as well as the cogency of any attempt to anchor the intrinsic value of nonhuman
species in the intrinsic value of nonhuman individuals. But for-
tunately for believers in the intrinsic value of species, the concept
of intrinsic value is broader than the concept of individual welfare, however interpreted. There is no contradiction in ascribing
intrinsic value to natural objects other than living individuals. At
the cost of forfeiting the clearest and least controversial model for
intrinsic value, that of the intrinsic value of human individuals,
species preservationists can attribute intrinsic value to species
themselves or to ecosystems. In the latter case the value of a species would be at least partly instrumental, but instrumental to a
source of value independent of human values—ecosystems. Such
theories would therefore avoid anthropocentrism.

I now turn to an examination of nonantrhopocentric approaches that do not use the analogy between human and nonhuman individuals. These theories must generate the value of species and/or ecosystems from other analogies or from none at all.

While consideration of such approaches involves venturing
into largely uncharted waters, a number of species preservationists have perceived the bankruptcy of individualism as a support
for their policies. Leopold seems to be among them, as the second
passage I quoted from him in section 8.2 defies individualistic
interpretation. After conceding that species and communities are
resources and will be used, he affirms their right to "continued ex-

istence in a natural state."[1] These comments seem to invoke a right of species to continued existence. Can we make sense of attributions of intrinsic value to species, in terms of species rights?

Rights are generated from interests or claims filtered in some manner by moral principles. From what interests or claims are rights of collectives to be generated? They cannot be generated simply from the interests of individual members because the interests of individuals will often conflict with procedures most likely to perpetuate the species, which must survive the death of its members. In this respect the attribution of rights to a species must differ from the attribution of rights to a corporation. At first glance rights attributed to corporations appear to provide an analogy for the extension of rights to a collective. But the purpose of a corporation is defined as the goal chosen by its shareholders (to make a profit, to further a cause, et cetera), and its value is instrumental to those shared goals of the owners. In this sense the rights of corporations can be generated from the rights of individuals—corporations are only legal fictions, instruments for the pursuit of goals held by shareholders as individuals. Attributions of rights to species as collectives, however, cannot be instrumental in this sense because they are introduced as an interpretation of claims concerning the intrinsic value of nonhuman species. What is in the interest of a species may prove directly contrary to the interests of particular members. I have shown that the perpetuation of a species may conflict with the interests of members individually, and this shows an essential disanology between rights of corporations and rights of species.

It might be suggested that rights of species are somehow generated from the interests of species, but the concept of interest of species is not at all clear. Individual lives are bounded by birth and death, and these parameters guide judgments of individual interest. Anything that threatens to cause the death of an individual is clearly against that individual's interest, other things being equal. But suppose a species is suffering a slow but steady decline

[1] Aldo Leopold, *A Sand County Almanac* (London: Oxford University Press, 1949), p. 204.

in population because of competitive pressure from another species (human or otherwise). Would it be in the interest of the species to come under steady adaptational pressure that both fuels its decline in population and, simultaneously, increases the likelihood that it will speciate before it becomes extinct? That is, there is no single manner in which the "life" of a species is terminated. Also, it is not obvious whether having a large population is in the interest of a species or not. Is a species "better off" if it has a larger population? Or is a small, robustly healthy population preferable? Is it better for a species to be moving toward more specialized adaptations, or are more opportunistic adaptations preferable from its point of view? It is not just that these questions have no ready answer. Rather, they seem to be odd questions as it is unclear whether or not they are to be answered in terms merely of the likelihood that the species will be perpetuated. Without guidance concerning the "interests" of species, there is no sense of how to proceed toward an answer to them.

These puzzles derive largely from the fact that interests and rights have been traditionally—semantically and in all other ways—treated as individualistic. To extend attributions of rights and morally relevant interests from human to nonhuman individuals involves no important alteration in moral concepts. While the concepts would be more broadly applied, their logic would remain intact, although, as we have seen, persuasive arguments militate against individualistic approaches to species preservation. To apply the concepts of rights and interests to nonhuman species as collectives is not only to expand the application of the concepts; it radically alters their very logic as no reasonable analogies exist for reconstructing them. We are left without any clear guidelines for deciding what rights a species has because one cannot generate such rights from interests in any meaningful sense of that term.

But this result undermines considerably the original motivation for appealing to rights. Since attributions of rights seem clear in some contexts, it was thought that interpreting claims that species have intrinsic value as claims that they have rights might help to clarify the vague concept of intrinsic value. Appeals to individ-

171

ual rights of members of nonhuman species, while retaining some hope to clear interpretation on a plausible theory of rights-holding, proved inadequate for concerns of species preservationists. But attributions of rights to collectives, while more likely to be adequate to preservationists' concerns, deviate from the logic of the traditional concept of rights so extensively as to undermine any significant analogies to human rights—the area where the concept of rights has some clarity. To say that a species has rights, then, is no more clear than to say that the species has intrinsic value. Once this is recognized, the violence done to linguistic tradition by speaking of rights of collectives rather than of individuals cannot be justified theoretically. It may be justified forensically, as has been noted, but it goes no way toward clarifying the difficult concept of intrinsic value.

Of course there are other ways of explaining and justifying attributions of intrinsic value to ecosystems and nonhuman species as collectives. J. Baird Callicott has offered two separate, and apparently quite different, approaches. In at least one he chooses the courageous course of building a positive, nonindividualistic theory accounting for the value of species and ecosystems. In the next two sections I will examine these two novel approaches.

9.2 *Human Sentiments*

Callicott, who has argued in several essays against interpreting the intrinsic value of species as deriving from individual welfare or concern, is convinced that nonhuman species have value not reducible to the value of individuals.[2] In his view the intrinsic value of nonhuman species is generated from a combination of David Hume's ethical principles and Charles Darwin's biological theories.[3] Callicott begins by asking why, if humans result from natural selection and survival of the fittest in a hostile world, they

---

[2] See, especially, "Animal Liberation: A Triangular Affair," *Environmental Ethics* 2 (1980): 311-338, and "On the Intrinsic Value of Nonhuman Species," in *The Preservation of Species*, ed. Bryan G. Norton (Princeton; N.J.: Princeton University Press, 1986).

[3] Callicott, "Intrinsic Value."

have any altruistic sentiments at all. Following Hume, he answers that human altruism, like egoism, is a primitive affection. He argues that, according to Hume, "self-love" and "sympathy" are on an equal footing as two primitive human moral sentiments. Hume forgoes concern with pleasure, pain, reason, or interests as *bases* for intrinsic value. Human sentiments are not derived from facts, rationally, but are human reactions, "*affects*," toward the "given" world.[4]

As such, these sentiments are not to be justified rationally but, at most, to be explained. Darwin's theory of natural selection, if allowed a group selection interpretation, can provide such an explanation. Rather than becoming progressively rapacious, selfish, and merciless in dealings with others (as the simplest picture of the workings of natural selection would suggest), the human race has developed altruistic sentiments. This was necessary because humans must survive a prolonged helpless period following birth in order to grow up to reproductive maturity. "Parental and filial affections" emerged in order to ensure the necessary care during this period and became the basis of altruistic motives. These developed into broader altruistic sentiments because affection and altruistic behavior extended through an entire tribe improved the survival chances of that tribe vis-à-vis other tribes that had narrower sentiments. And an impetus existed to spread these sentiments still more broadly because larger tribes had a natural advantage in intertribal competition.

These sentiments, then, have a natural, evolutionary origin, provided one accepts that natural selection applies to groups as well as individuals. Callicott continues:

> To whom or what these affections are directed, however, is an open matter, a matter of cognitive representation—of "nurture" not "nature." A person whose social and intellectual horizons are more or less narrow regards only a more or less limited set of persons and a more or less local social whole to be intrinsically valuable. To perceive non-human

[4] Also see E. O. Wilson, *Biophilia* (Cambridge, Mass.: Harvard University Press, 1984).

species as intrinsically valuable involves, thus, not only the moral sentiments, but an expansion of the *cognitive representation* of nature.[5]

At this point, perhaps, something like Taylor's "biocentric outlook on nature" could affect a person's cognitive representation and encourage a broader set of social sentiments, including a sentiment attributing intrinsic value to nonhuman species.[6] Indeed, it might be thought of as providing one model for how mere felt preferences could be replaced by considered ones within an altered world view. This approach, then, provides an explanation of the origins of the view, held by many environmentalists, that nonhuman species have intrinsic value. As such, it is important because it offers an explanation of the origins of a hitherto poorly explained sentiment or idea.

The first objection to Callicott's novel approach might be that it is not all that novel compared to the approaches previously examined and rejected. Surely the primitive affections deriving initially from the parent's natural affection for the child are directed at individuals. And, while Callicott has abandoned the attempt to justify such sentiments, thereby fulfilling his promise not to support the rights of species with individual interests, he must somehow explain how concern for individual human children undergoes a metamorphosis into concern for nonhuman species. He must also explain how concern for species so derived can avoid conflict with the more basic concern for individuals that haunted the efforts rejected in Chapter 8. It appears, then, that Callicott's biosentimentalism does not really escape the damaging individualistic bias detailed above.

Leaving this point aside, since Callicott's approach explains attributions of intrinsic value without attempting to justify them, it therefore suffers from the standard objection to Humean ethics. Most philosophers reject Hume's account because it is perceived to be relativistic and offers no basis for objective argument when

[5] Callicott, "Intrinsic Value."

[6] See Paul W. Taylor, "The Ethics of Respect for Nature," *Environmental Ethics* 3 (1981): 197-218. Also, see Chapter 8 above.

disagreement arises. Callicott provides an answer to this charge but only a partial one. He points out that a "consensus of feeling" emerges on important points, providing a "functional equivalent of objective moral truths." There is a standardized "psychological profile" on such matters as murder and theft that transcends cultural differences. Murderers and thieves don't just have somewhat different sentiments from most people; they are viewed in all cultures as "freakish" and beyond moral acceptability. Further, such standardizations of conduct are explained in Darwinian terms—they may be fixed by natural selection.[7]

These points, while well taken, go only so far. They explain how we come to have and share the sentiments we do. But they do not explain how we come to have our "intellectual horizons." As current debates about endangered species policy show, there are no shared beliefs about how widely the moral sentiments should be applied. There is no basis to argue with the anthropocentrist here. Either someone has altruistic sentiments directed broadly at all species, or he does not. If not, there is no basis for rational disagreement or argument designed to convince him to broaden these sentiments. Callicott's account provides an explanation of attributions of intrinsic value but only at the very high cost of placing those attributions beyond moral justification.

If a majority of Americans shared the sentiments of altruism toward species, this approach would support a preservationist policy. If not, it would be no help—the question reduces to actual attitudes. But a majority of people, even a majority of environmentalists, seem to favor instrumental justifications for preserving species. The majorities no doubt increase dramatically as one moves down the phylogenetic scale from mammals and vertebrates to invertebrates and plants.[8] So, while Callicott's account provides an explanation of an important (probably minority) attitude, it provides no firm basis for arguments in favor of a general policy of species preservation, so long as the views explained remain minority views.

[7] Callicott, "Intrinsic Value."
[8] On this point, see Stephen R. Kellert, "Social and Perceptual Factors in the Preservation of Animal Species," in *The Preservation of Species*.

9.3 Holism

Drawing on one interpretation of Leopold's land ethic, Callicott has also developed another basis for attributing intrinsic value to nature. If species can be seen as occupying logical ground *between* individuals and ecosystems—they are ever-changing collections of individuals that at any point in time are components of functioning ecosystems—then deriving the value of species from ecosystems, rather than from individuals, may be a viable possibility. Callicott explores this ethical system as follows.

A society is constituted by its members, an organic body by its cells, and the ecosystem by the plants, animals, minerals, fluids, and gases which compose it. One cannot affect a system as a whole without affecting at least some of its components. An environmental ethic which takes as its *summum bonum* the integrity, stability, and beauty of the biotic community is not conferring moral standing on something *else* besides plants, animals, soils, and waters. Rather, the former, the good of the community as a whole, serves as a standard for the assessment of the relative value and relative ordering of its constitutive parts and therefore provides a means of adjudicating the often mutually contradictory demands of the parts considered separately for *equal* consideration. If diversity does indeed contribute to stability (a classical "law" of ecology), then *specimens* of rare and endangered species, for example, have a *prima facie* claim to preferential consideration from the perspective of the land ethic. Animals of those species, which, like the honey bee, function in ways critically important to the economy of nature, moreover, would be granted a greater claim to moral attention than psychologically more complex and sensitive ones, say, rabbits and moles, which seem to be plentiful, globally distributed, reproductively efficient, and only routinely integrated into the natural economy.[9]

[9] Callicott, "Animal Liberation," pp. 324-325.

This holistic ethic ascribes value to species because species are essential to the continued functioning of organic systems. If these systems are attributed *intrinsic* value, then there would be a powerful reason for protecting species deriving from their instrumental value in promoting the ultimate good, ecosystems. While species have no intrinsic value according to this approach, the system of value described is nonanthropocentric, and species have value independent of their value to humans.

There is, of course, a danger that anthropocentrism could steal back into this system if the value of ecosystems is ultimately envisaged as benefit to humans, as was discussed in Chapter 3 and Chapter 4. If the proposed "land ethic" is to succeed as a truly nonanthropocentric possibility, then the value of ecosystems must not reduce to the value humans derive from them. This is not to suggest that they cannot have, concurrently, instrumental value to humans, only that their value must exceed their value to humans. If the view that ecosystems are intrinsically valuable could be defended, the result would be a land ethic that transfers value to species, not to individuals. In the process it would justify special treatment of individuals who are members of endangered species. Extinct species can no longer contribute to the intrinsic, holistic value of ecosystems.

A holistic land ethic could exist in two versions, monistic and pluralistic. Monistic holism attributes intrinsic value only to organic communities; pluralistic holism recognizes intrinsic value in human and nonhuman individuals as well as in ecosystems. To recognize multiple sources of intrinsic value is, of course, to admit the possibility of conflict among these sources, with the need for hard moral choices to balance competing claims. As was argued in the last chapter, the interests of individual nonhumans are often in conflict with efforts at ecosystem and species preservation, so the pluralistic alternative faces special problems.

Perhaps Callicott recognizes these problems, as he opts, without explanation or discussion, for monistic holism: "the land ethic manifestly does not accord equal moral worth to each and every member of the biotic community; the moral worth of individuals (including, n.b., human individuals) is relative, to be as-

177

sessed in accordance with the particular relation of each to the collective entity which Leopold called 'land.' "[10]

But monistic holism has startling implications for policy. Monistic holism seems committed to valuing humans only insofar as they contribute to ecosystem functioning. Assuming, as most environmentalists now do, that human activities have on the whole a negative effect on ecosystem functioning, individual human lives would be negatively valued. A member of an ecologically important endangered species would be worth many human lives. That monistic holism has this consequence is illustrated when Callicott quotes Edward Abby, who said he would sooner shoot a man than a snake, as a representative of this land ethic.[11]

Monistic holism is, then, based on a highly unusual and sure to be widely rejected system of ethics. Even if a small number of radical environmentalists can accept these *antihumanistic* consequences, there will never, in the foreseeable future, be a societal consensus favoring such an extreme view. Monistic holism cannot be enlisted as an important component of any rationale for preserving species. The values on which it rests are too idiosyncratic to carry any weight politically.

A pluralistic version of holism may be more palatable. According to such a view, individuals *and* systems can *both* be thought to have intrinsic value. A member of an endangered plant species would be instrumentally valuable for the contribution it makes to systems that are intrinsically valuable; but human individuals and some or all nonhuman individuals could, likewise, be considered intrinsically valuable. This position seems, initially, similar to Paul Taylor's views as previously discussed. But Taylor attributes intrinsic value to species and ecosystems because they are collections of intrinsically valuable individuals.[12] Callicott claims that ecosystems have intrinsic value independently of participating individuals, assuming that any value ascribed to individuals is a function of their contribution to the valuable whole of which they are a part. I am suggesting a third possibility—that some or all

[10] Ibid., p. 327.
[11] Ibid., p. 326.
[12] Taylor, "Ethics of Respect," pp. 197-198.

individuals have intrinsic value and that some or all ecosystems, considered as wholes, have value independent of individual values.

As far as I know, no author has begun the difficult task of developing pluralistic holism as a value theory. It has some promise as a means of justifying fairly widespread intuitions about the importance of other species without assigning negative value to human individuals. Such a system would not have a unified theory of value, and it would be very difficult to work out rules for assessing *comparative* values: Could several ecosystems outweigh the value of one human baby? It is not just that one does not know what answer to give to this question; one has no idea how to begin to answer it. The problem is more severe than that faced by Taylor, who must develop a system of ethics on which one can ascribe equal consideration to the interests of nonhuman as well as human individuals. Of this task Taylor can say: "The impartiality demanded of us is no different than the impartiality we must exercise when we are trying to decide what is the fair thing to do when our personal interests are in conflict with those of our fellow humans."[13] He can say this because he believes that individual interests of human and nonhuman individuals, however different in weight, are commensurate. They can be placed on a single scale of individual welfare. This option is not readily available to my pluralist. The value of individuals and the value of ecosystems originate in distinct values, and the latter is not reducible to questions of individual welfare. There is, then, no straightforward way to make such values commensurate.

Until such a theory has been developed in detail, it would be highly speculative to enlist it as a basis for species preservation. In general it can be said that however interesting are holistic theories of value, whether monistic or pluralistic, they remain far too undeveloped to serve as a basis for policy formulation at this time. More research and discussion of these options should be a priority among environmental ethicists.

[13] Paul W. Taylor, "In Defense of Biocentrism," *Environmental Ethics* 5 (1983): 243

9.4 Prospects

The passion with which species preservationists attribute intrinsic value to nature is more easily explained than justified. Besides expressing their moral outrage at human mistreatment of nature's wonders, such attributions anchor that moral outrage in fashionable ethical principles and promise to remove any sense that the obligation to protect species is accepted optionally, if at all. But much of the appeal of such attributions derives, either explicitly or implicity, from the unwarranted assumption that they would automatically link preservationist concerns to the highly developed and relatively noncontroversial ethics for protecting interests and rights of human individuals.

If one rejects the individualistic route and the unwarranted assumptions that go with it, attributions of intrinsic value to nonhuman species are cast adrift, lacking any clear and uncontroversial foundation in value theory. The defender of such attributions must, therefore, first develop, clarify, and justify an axiology that recognizes intrinsic values transcending concerns for individual welfare and only then begin to argue that nonhuman species have the characteristics constitutive of such value. Until these very difficult tasks are undertaken by defenders of nonanthropocentrism, it is difficult to criticize their often casual remarks about the intrinsic value of nonhuman species. I have argued mainly that these remarks can borrow no plausibility from association with the uncontroversial case of intrinsic value attributed to human individuals. Once this point is accepted, the popular method of negative analogical reasoning from the human individual case loses force. The nonanthropocentrist must take a positive approach, clarifying the conception of intrinsic value employed before justifying its application to species. This is a monumental task not yet begun. Not begun, perhaps, because the sirens of individualism inexorably seduce the nonanthropocentrist.

I do not intend to discourage nonanthropocentrists from attempting to clarify and justify their arguments. On the contrary, the main thrust of my remarks has been to demonstrate that most attempts to this point have been misguided. One effect of my arguments may be to encourage nonanthropocentrists to eschew individualism and develop a theory of natural value derived neither from human individual interests nor built upon analogy to them.

To see how far nonanthropocentrists have to go in their task, one need only quote Tom Regan's much-discussed appeal for a nonanthropocentric value theory:

> Two questions which I have not endeavored to answer are: (a) what, if anything in general, makes something inherently good, and (b) how can we know, if we can, what things are inherently good? The two questions are not unrelated. If we could establish that there is something (X) such that, whenever any object (Y) has X it is inherently good, we could then go on to try to establish how we can know that any object has X. Unfortunately, I now have very little to say about these questions, and what little I do have to say concerns only how not to answer them.[14]

To Regan's credit, he recognizes clearly the task at hand. And he does not opt for the easy way out—resorting to the indirect method of merely denying disanalogies. But his remarks are worrisomely tentative and abstract. When a leading advocate of intrinsic value of nonhuman species has so little to say concerning what he means by his views or how he would justify them in positive terms, one cannot but be skeptical about their role in public debate concerning policies having palpable effects on economic efficiency and other human values.

No committed species preservationist can resist the attraction of an argument for protecting species that is independent of the contingent and ephemeral arguments derived from human uses, tastes, and fashions. But policy formation cannot be based upon whim, however well intentioned. Arguments for species preservation based on attributions of intrinsic value to nonhuman species cannot play a decisive role in formulating public policy until

[14] Tom Regan, "The Nature and Possibility of an Environmental Ethic," *Environmental Ethics* 3 (1981): 33.

they are developed, clarified, and justified in a clear and explicit manner. Only then can they have persuasive force and only then can they assume a role in serious policy debates about what, if anything, ought to be done to preserve other species.

Nonanthropocentrism remains, then, an interesting and important philosophical theory. Perhaps soon nonanthropocentrists will reject individualism as a false start and turn to the real task at hand—to develop an axiology that supports concern for species and ecosystems rather than individuals. The development and justification of philosophical theories is a slow process, however. Meanwhile, concerned scientists predict that as many as one quarter of all species will be lost in the next quarter century. It is an exercise in monumental understatement to remark that philosophical clarification of appeals to intrinsic value of nonhuman species may come too late.

C

TRANSFORMATIVE VALUES AND SPECIES PRESERVATION

T E N

TRANSFORMATIVE VALUES

10.1 *The Situation*

I began by defining four categories of reasons for protecting species. These categories, defined in terms of the values to which they appeal, were formed by the intersection of two dichotomies, one distinguishing values according to who or what is served by them and the other according to the type of value involved. Anthropocentric arguments comprehend all reasons that ultimately appeal to the intrinsic value located in human beings, while nonanthropocentric arguments appeal to intrinsic value located in nonhuman species. Cutting across this dichotomy is another one: the value of an experience can lie either in its fulfilling an existing, prior preference of some individual (demand value) or in its altering or transforming such preferences.

Arguments available to support species protection can then be classified as appealing to: (a) human demand values, (b) human transformative values, (c) nonhuman demand values, or (d) nonhuman transformative values. To countenance arguments based on (d) would require an assault on a very widely accepted metaphysical dichotomy, which singles out humans as the only known species whose members can rationally consider and reject value commitments. I have chosen not to give serious consideration to values in this category, as my purpose is to examine and then establish viable bases for policy formation.

Serious consideration is being given, then, to values in three categories. No one, including those who place no special stock in species preservation, seriously denies that nonhuman species satisfy human demand values. Part A, while not attempting to ex-

185

haust the sources of such satisfactions, emphasizes some easily overlooked instances, concluding that nonhuman species and the ecosystems they compose serve human demand values in many not readily understood or quantified ways. Yet species protectionists are justifiably uneasy with resting their case wholly on the positive manner in which nonhuman species satisfy human demand values.

Most preservationists who have gone beyond human demand values in their pleas for enlightened policies have appealed to values entirely independent of human objectives. This nonanthropocentric alternative is ethically radical, espousing as it does new moral demands on policy formation—demands originating outside of and capable of competing with human demands and yet it has attracted numerous and vocal proponents. It expresses in moral terms the outrage of environmentalists at human destruction of other organisms and attempts to show the inherent limits of human exploitation of nature.

But the initial attraction of this approach proved, in Part B, illusive. The greater portion of the attraction derives from the promise to annex the well-developed and largely uncontroversial system of ethical rules governing treatment of human individuals to the cause of species preservation. But attempts to move, by analogy, from moral concern for human individuals to a corresponding moral concern capable of supporting a policy of species preservation face two serious disadvantages. First, species preservationists require a basis for protecting a wide range of species spread throughout the phylogenetic scale. To attain this comprehensive basis, they must extend the analogy from human individuals to all individual organisms. This cannot but weaken the analogy, as it taxes credibility to deny any difference, morally, between "mistreating" plants and mistreating humans.

Second, even leaving aside these problems of scope, concern for individual specimens of nonhuman species supports policies often at odds with the best strategy for preserving species. A species is a composite of individuals surviving beyond the death of any of its individual members, and there is no necessary connection between the protection of individual interests and the preservation of species into indefinite time.

Locating intrinsic value in species themselves or in ecosystems holds some promise of providing nonanthropocentric reasons for preserving species, reasons not dependent upon individual concerns. But these approaches go beyond the simple annexation of standard ethical rules and principles; they require the development of an axiology clarifying and justifying their claims. No uncontroversial bases for analogy are available here, and the task of nonindividualistic nonanthropocentrists is an arduous one. They must define a nonindividualistic conception of intrinsic value and then state some positive characteristic standing as the mark of such value. Only then can they begin to argue that nonhuman species and ecosystems have the relevant characteristic and to derive policies from those values.

Thus, while no arguments rule out appeals to nonindividualistic intrinsic value, the development of an adequate ethic based upon such a foundation is surely a long way off. Even if promising alternatives were being suggested at this point, it is dubious that they could receive the necessary clarification and support to inform environmental policy in time to avoid the cataclysmic reduction in biological diversity feared by scientific experts.

Here, then, is the dilemma: Ought species preservationists to continue espousing nonanthropocentric reasons for species preservation, knowing full well that these reasons cannot yet be supported by a clear and rationally defensible axiology, or should they fall back on unquestioned human demand values as the full basis for the policies they recommend? The first alternative supports obligations sufficiently strong to limit human destruction of other species but at the cost of leaving preservationists open to the charge that their arguments are based on unclear and unsupported value premises. The second alternative avoids this charge but only at the cost of putting the goals of species preservationists on a par with other human demands. How can preservationists argue that the long-term demand value of species should receive high priority in a world faced on every hand with pressing and immediate human demands for food, shelter, and other basic needs?

It is the central thesis of this book that this excruciating dilemma is a false one because it unnecessarily contracts the range of human values to those founded on demands for given preferences. However important are the ways species serve human demands, there exists an entirely different category of important human values that cannot themselves be reduced to these demands. To assume that policies favoring species preservation must either be based on intrinsic value of nonhumans or else that these policies must rest on no sounder basis than the shifting, contingent preferences humans express for material goods and services is to ignore the role of other species and varied ecosystems in forming and transforming values. The next subject requiring extensive discussion is, then, transformative values.

10.2 Transformative Values

The way out of the preservationist's perplexing dilemma is to refuse to accept the view that all felt preferences are on a par. Some can survive a rigorous process of examination and emerge as considered preferences; others cannot. In this way species preservationists can give importance to some demand values while criticizing and rejecting others as less worthy of concern. They can accept the prima facie value of satisfying individual human preferences while reserving the right to criticize some of these preferences as overly consumptive, materialistic, and unworthy of satisfaction.

This crucial distinction between worthy and unworthy preferences requires that we recognize an ambiguity in the concept of rationality as applied to human ends and objectives. To say that someone is a rational seeker of ends is sometimes to say no more than that she is *capable of* countenancing reasons in deciding what ends to pursue. At other times it is to say that she has *correctly identified and weighted* the factors that justify pursuing a chosen goal or goals. Only demand values that are rational in the second, stronger sense have an unquestioned claim in policy formation, and only those need be invoked to support species preservation for its benefits to human consumers. And at the same time this structure provides a role for a new category of values—those that I have called transformative.

A value system that includes transformative as well as demand values can limit and sort demand values according to their legitimacy within a rational world view. To the extent that one values having a rational set of felt preferences, experiences that contribute to the formation of a rational world view and an attendant adjustment of felt preferences have transformative value.

This more complex, though still anthropocentric, value system is doubly congenial to the goals of environmental preservationists. It allows them to express their legitimate concern that runaway expansion of human demand values, especially overly materialistic and consumptive ones, constitutes much of the problem of species endangerment. It also highlights the value of wild species and undisturbed ecosystems as occasions for experiences that alter those very felt preferences. Occasions such as these have value of a different order, although value still couched in human terms. Insofar as environmentalists believe that experience of nature is a necessary condition for developing a consistent and rational world view, one that fully recognizes man's place as a highly evolved animal whose existence depends upon other species and functioning ecosystems, they also believe that such experiences have transformative value. Experience of nature can promote questioning and rejection of overly materialistic and consumptive felt preferences. Appeals to the transformative value of wild species and undisturbed ecosystems thereby provide the means to criticize and limit demand values that threaten to destroy those species and ecosystems while at the same time introducing an important value that humans should place upon them.

An illustration may prove helpful here. Suppose an adult comes upon a child playing in the woods. The child is gleefully destroying eggs from the nests of groundbirds. The adult gently explains to the child that eggs are necessary to hatch baby birds and shows the child baby birds in another nest. The child is fascinated, watches the baby birds being fed by the mother, and loses interest in his destructive game. Now he begins to show solicitous concern for the welfare of birds and asks many questions. Eventually, he grows up to be an amateur ornithologist, deriving untold pleasure from a lifetime interest in birds. The initial appeal of the destructive felt preference and the demand value represented has now been transformed. To the extent that one believes the child's posteducational preferences are more rational and less open to criticism, one also believes that the encounter with the birds has value: it has transformed irrational and indefensible felt preferences into more defensible considered preferences.

This example stands as a microcosm of the power of nature to transform values across society. When the initial felt preferences of a whole society are askew, the need for transformative value is correspondingly more urgent, as Joseph Sax explains in his compelling book, *Mountains without Handrails*:

> The preservationist is not an elitist who wants to exclude others, notwithstanding popular opinion to the contrary; he is a moralist who wants to convert them. He is concerned about what other people do in the parks not because he is unaware of the diversity of taste in the society, but because he views certain kinds of activity as calculated to undermine the attitudes he believes the parks can, and should, encourage. He sees mountain climbing as promoting self-reliance, for example, whereas "climbing" in an electrified tramway is perceived as a passive and dependent activity.[1]

Sax perceives that preservationists are moralists: they attempt to transform values. He also sees that the preservationists' moralism extends to the whole society. They are extolling the virtues of a society that protects natural places and the values they stand for:

> The preservationists do not merely aspire to persuade individuals how to conduct their personal lives. . . . The parks are, after all, public institutions which belong to everyone,

[1] Joseph Sax, *Mountains without Handrails* (Ann Arbor: University of Michigan Press, 1980), p. 14.

not just to wilderness hikers. The weight of the preservationist view, therefore, turns not only on its persuasiveness for the individual as such, but also on its ability to garner the support—or at least the tolerance—of citizens in a democratic society to bring the preservationist vision into operation as official policy.[2]

The case being made extends to societal values: What type of society shall we have? An answer to this question requires, Sax believes, appeals to values that can only be found in, and protected in, nature. This is the transformative value of nature.

10.3 The Transcendentalist Tradition

Appeals to transformative values are not without intellectual antecedents. Contemporary concern for the protection of nature can be linked to the venerable American intellectual tradition deriving from the transcendentalist movement led by Emerson and Thoreau. Emerson and Thoreau believed that experience of nature elevates human values. Such experiences illuminate, through analogy and metaphor, transcendent, spiritual values. They serve, in the transcendentalists' world view, a function parallel to the role here proposed for nature as a transformative agent in human value formation and alteration. Transcendentalism thus provides an intellectual framework for environmentalism, an old wineskin into which environmentalists can pour new wine.

The framework to which I refer is nicely illustrated by Thoreau, when he says, "One value even of the smallest well is, that when you look into it you see the earth is not continent but insular. This is as important as that it keeps butter cool."[3] This brief passage shows how a natural (though here human-altered) phenomenon both fulfills demand values and instructs humans as to their "place" in the cosmos. He continues:

[2] Ibid., p. 103; also see George R. Hall, "Conservation as a Public Policy Goal," in *Politics, Policy, and Natural Resources*, ed. Dennis L. Thompson (New York: The Free Press, 1972), p. 186.
[3] Henry David Thoreau, *Walden* (New York: The New American Library, 1960), p. 63.

When I looked across the pond from this peak toward the Sudbury meadows, which in time of flood I distinguished elevated perhaps by a mirage in their seething valley, like a coin in a basin, all the earth beyond the pond appeared like a thin crust insulated and floated even by this small sheet of intervening water, and I was reminded that this on which I dwelt was but *dry land*. . . .

Both time and place were changed, and I dwelt nearer to those parts of the universe and to those eras in history which have most attracted me. . . . Such was the part of creation where I had squatted;—

> "There was a shepherd that did live,
> And held his thoughts as high
> As were the mounts whereon his flocks
> Did hourly feed him by."

What should we think of the shepherd's life if his flocks always wandered to higher pastures than his thoughts?[4]

For the transcendentalists, experiences of natural objects are occasions for the adjustment of one's thoughts about one's place in the great system encompassing space and time and, simultaneously, occasions for "higher thoughts." Environmentalists can learn from this approach: their concern that human values are unacceptably consumptive and materialistic is shared by the transcendentalists, and both groups share the belief that through experiences of nature, a new sense of value emerges.

While some modern environmentalists have perceived their debt to the transcendentalists, they have been embarrassed by two features of the system of Thoreau and Emerson and have failed to embrace them as the patron saints of the movement. First, environmentalists have been uncomfortable with the spiritualistic values espoused by the transcendentalists. Nature provides a visible means to discover invisible values—but there is no independent means to ascertain or understand these values. As-

[4] Ibid., pp. 63-64.

sertions of such values are untestable, except by subjective and nonrational "intuitions" of the spirit.[5]

Second, it is not clear how much protection transcendentalism actually offers unspoiled nature against the inroads of human technology. Although the locomotive became, in the nineteenth century, the symbol of materialism, urbanization, and the intrusion of technology into the garden of nature,[6] Emerson was capable of exulting in it as an expression of human creativity. Hence we find him offering this paean to technology:

> The useful arts are reproductions or new combinations by the wit of man, of the same natural benefactors. He no longer waits for favoring gales, but by means of steam, he realizes the fable of Aeolus's bag, and carries the two and thirty winds in the boiler of his boat. To diminish friction, he paves the road with iron bars, and mounting a coach with a shipload of men, animals, and merchandise behind him, he darts through the country, from town to town, like an eagle or a swallow through the air. By the aggregate of these aids, how is the face of the world changed, from the era of Noah to the era of Napoleon![7]

Thoreau, on the other hand, took a disdainful attitude toward this technological marvel, likening the menacing locomotive's whistle to "the scream of a hawk sailing over some farmer's yard, informing me that many restless city merchants are arriving within the circle of the town."[8] Thus there seems to be an ambivalence in the transcendentalist view of nature and technology, which can be explained by the transcendentalists' ontology:

[5] See Frederick Ives Carpenter, "Transcendentalism," in *American Transcendentalism: An Anthology of Criticism*, ed. Brian Barbour (Notre Dame, Ind.: University of Notre Dame Press, 1973), pp. 24-25.

[6] See Leo Marx, *The Machine in the Garden* (New York: Oxford University Press, 1964), especially Chapter 1. Marx mentions several examples from Hawthorne, Wordsworth, etc. Also see Thoreau, *Walden*, pp. 67, 82.

[7] Ralph Waldo Emerson, "Nature," in *Selected Essays*, ed. Larzer Ziff (New York: Penguin Books, 1982), p. 41.

[8] Thoreau, *Walden*, p. 82.

Philosophically considered, the universe is composed of Nature and the Soul. Strictly speaking, therefore, all that is separate from us, all which Philosophy distinguishes as *not me*, that is, both nature and art, all other men and my own body, must be ranked under this name, Nature. In enumerating the values of nature and casting up their sum, I shall use the word in both senses—in its common and in its philosophical import. In inquiries so general as our present one, the inaccuracy is not material; no confusion of thought will occur. *Nature*, in the common sense, refers to the essences unchanged by man; space, the air, the river, the leaf. *Art* is applied to the mixture of his will with the same things, as in a house, a canal, a statue, a picture. But his operations taken together are so insignificant, a little chipping, baking, polishing, and washing, that in an impression so grand as that of the world on the human mind, they do not vary the result.[9]

Ontologically, transcendentalists sharply separate the human self and spirit from the body and the natural world. Nature, as man learns from it, is not embodied in concrete, particular objects but in "essences" that are unalterable by human activity. Epistemologically, then, learning from nature depends not upon the senses and upon scientific observation but upon super-rational intuitions not subject to discussion and debate.[10] The values thus served are other-worldly, spiritual, and often seem cut loose from day-to-day activities.[11] These factors explain the malleability of

[9] Emerson, "Nature," p. 36.

[10] Carpenter, "Transcendentalism," pp. 24-25.

[11] This latter indictment seems unfair when applied to Thoreau, who was much less explicit than Emerson about the philosophical principles on which his system rests. See, for example, his very down-to-earth evaluative discussion of the activities of tree-cutters, where, of a chopper with whom he was acquainted, he said: "He was a skillful chopper, and indulged in some flourishes and ornaments in his art. He cut trees level and close to the ground, that the sprouts which came up afterward might be more vigorous and a sled might slide over the stumps . . ." (p. 101). These values can hardly be identified with the abstract, spiritual ones characteristic of Emerson's full-bloom transcendentalism. Thoreau's instincts in this direction explain a growing reluctance to equate his views with those of Emerson. See, for example, Joseph W. Krutch, *Henry David Thoreau* (Westport, Conn.: Greenwood Press, 1948), pp. 46-51.

the transcendentalists' views on technology, material wealth, and the alteration of nature by man. Human artifice exists in the physical world. True value exists in a separate, spiritual world. One can as well learn from nature altered by man as from pristine wilderness because the values learned are abstract, spiritual, and independent of human alteration.

So, with good reason, environmentalists have shied away from a strong and unequivocal endorsement of the transcendentalist world view and its system of value. But these apprehensions should not obscure the similarity in structure between transcendentalism and the approach to valuing nature I am here trying to explain. The transcendentalists attributed no intrinsic value to nature independent of human consciousness or evaluations. Emerson is quite clear that humans *use* nature for various human purposes, the highest to serve as a means for achieving spiritual enlightenment. He says: "Nature is thoroughly mediate. It is made to serve. It receives the dominion of man as meekly as the ass on which the Savior rode. It offers all its kingdoms to man as the raw material which he may mould into what is useful."[12]

Nature thus alters human values but not by embodying an alternative locus of intrinsic value. Insofar as nature has value for the transcendentalists it merely expresses a "higher" value found in immaterial essences. But, for the transcendentalists, experience of nature transforms human value systems, undermines materialism, elevates our values beyond day-to-day cares of making a living and gaining wealth. This multileveled structure in transcendentalist thought, the sense that nature, in addition to fulfilling demand values (nature used as commodity), also serves a higher function—to alter, enlighten, and improve human value systems—can be abstracted from their system and given new life in the present context. Modern environmentalists cannot, of course, adopt the transcendentalist system intact. It is vague in its key terms and is entirely lacking an effective epistemology. My point is only that the transcendentalist value system, by emphasizing higher ideals in human values and by ascribing an important role

[12] Emerson, "Nature," pp. 57-58.

to experiences of wild nature in the pursuit of those goals, provides a general structure of anthropocentric values capable of providing a blueprint for developing a modern, and more adequate, conception of environmental values.

This structure in less developed form has existed as a powerful idea in the environmental movement from its inception. For example, John Muir's views on value in nature are often assumed to rely upon attributions of intrinsic value to nature, and Muir certainly believed that nature had value independent of humanity.[13] But there exists, side-by-side with that strain in Muir's thought, another value ascribed to nature:

> If you are traveling for health, play truant to doctors and friends, fill your pockets with biscuits, and hide in the hills of the Hollow, lave in its waters, tan in its gold, bask in its flower-shine, and your baptisms will make you a new creature indeed. Or, choked in the sediment of society, so tired of the world, here will your hard doubts disappear, your carnal incrustations melt off, and your soul breathe deep and free in god's shoreless atmosphere of beauty and love.[14]

This is the aspect of Muir's thought that led the journalist Ray Stannard Baker to describe Muir's views as "a complete expression of a deep human instinct which we have often felt, and throttled—the instinct which urges us to throw off our besieging restraints and complexities, to climb the hills and lie down under the trees, to be simple and natural."[15] It is this fact of environmental thought that I am suggesting can best serve as the blueprint for a powerful argument for preserving species.

10.4 The Embodiment of Culture Argument

Using a similar value structure, Mark Sagoff has developed a much-discussed argument, which he calls an aesthetic one, for

[13] See Stephen Fox, *John Muir and His Legacy: The American Conservation Movement* (Boston: Little, Brown, and Company, 1981), p. 52. Fox quotes from Muir's *A Thousand-Mile Walk to the Gulf* (Dunwoody, Ga.: Norman S. Berg, Publisher, by arrangement with Houghton Mifflin Company, 1916), p. 98.

[14] Muir, *Thousand-Mile Walk*, pp. 210-211.

[15] Quoted in Fox, *John Muir*, p. 118.

preserving natural objects as embodiments of American culture.[16] It will be instructive to examine Sagoff's argument here, adopting his somewhat nonstandard conception of aesthetics. Sagoff understands aesthetic qualities very broadly: "[a]n aesthetic quality is any quality named in a metaphorical way. The distinction between the non-aesthetic and the aesthetic and the distinction between the literal and the metaphorical coincide."[17] If objects have a quality metaphorically, they *express* or *exemplify* it. Objects can therefore serve as examples or "paradigms" of the qualities they express. These paradigms have a cognitive function: they provide "samples" by which we learn to recognize objects as having certain qualities. But they also shape the understanding of the quality in question. If the paradigm representing a quality changes, different objects will be perceived as having that quality.[18]

Sagoff begins by arguing that a purely utilitarian ethical system will not adequately preserve nature (for reasons similar to those given in Section 6.4).[19] Insofar as human preferences are malleable and only contingently dependent upon natural objects to fulfill them, they will present at best a shifting and unreliable basis for preservation.

Sagoff's positive position is characterized by two central features:

(1) the recognition that aesthetic experience not only fulfills fixed, unchanging desires but also shapes those desires by giving precise meaning to cultural ideals; and
(2) linkage of the idea-shaping function of aesthetic experience to particular features of specifically American historical patterns and cultural events.

These two features combine to provide a powerful and persuasive rationale for the protection of wilderness and wild species in America, by Americans. But the two features are separable: it is possible to accept the first while rejecting the second. I will ex-

---

[16] Mark Sagoff, "On Preserving the Natural Environment," *Yale Law Journal* 81 (1974): 205-267.
[17] Ibid., p. 248.
[18] Ibid., pp. 228-229.
[19] Ibid., pp. 206f.

plain, develop, and justify Sagoff's first feature in the present section, and I will advance reasons to question and at least modify the second feature in Section 10.5.

American culture values freedom. But freedom is not an inherently clear concept with a well-defined field of application that is necessarily fixed through time. Paradigms make certain important similarities and differences among things more conspicuous.[20] If one changes the paradigm that expresses freedom, one changes the very understanding of what it is to be free.[21]

We thus recognize the cultural value of art: "The business of the arts is to provide expressive objects and to represent other objects as expressive; therefore, art objects are themselves paradigms of aesthetic qualities and they represent other objects as paradigms."[22]

The destruction of a magnificent natural environment is, therefore, wrong for the same reason that the destruction of a great work of art is wrong: "In losing either, we lose the best example we have of a quality which we do not otherwise fully understand or on which we have no better grasp. The destruction of symbols is a step toward ignorance of the qualities those symbols express."[23]

The unusual cultural history of America, "Nature's Nation" as the American cultural historian Perry Miller described it, gives context to our interactions with natural environments and other species in the wild.[24] At the same time the importance accorded natural environments and other species has shaped that culture as well. The fact that Americans have identified wild things with freedom and independence has given a particular meaning to the American value of freedom. This meaning can be shared only by other nations that have shared a similar national experience. Americans came to associate their new-world virtues—independence, honesty, self-reliance—with the world of nature. These val-

[20] Ibid., p. 256.
[21] Ibid., p. 248.
[22] Ibid.
[23] Ibid., p. 259.
[24] Perry Miller, *Nature's Nation* (Cambridge, Mass.: Belknap-Press of Harvard University Press, 1967). And see Sagoff, "On Preserving the Natural Environment, p. 263.

ues were seen in contrast to the values of the decadent old-world cultures taken to be symbolized by cathedrals, great buildings, and mercantilist cities. The American values were exemplified, and thus shaped, by a feeling of affinity for wildness: the eagle soaring in the open sky, deer and bear running free in undisturbed woods, fending for themselves, making a life in the wilderness, much as the pioneering Americans chose to do.

This, according to Sagoff, furnishes the real reason for preserving wild places and wild species. They are symbols of the American cultural heritage and give meaning and value to it. To allow these species to become extinct would be to admit that the values they symbolize are no longer held in esteem by Americans.[25] It might well amount to embracing a new sense of freedom: not freedom from tyranny but freedom from hard work. Such a cultural value may be better symbolized by pop-up toasters and food processors than by the bear, the deer, or the eagle.

The argument implicit in this account of the value of other species exploits two crucial premises corresponding to the two features of Sagoff's account. First, the argument employs an empirical premise: if Americans treat their natural heritage with contempt, allowing other species to become extinct and pristine natural areas to become degraded, their values will change. Contemptuous treatment of the symbols of the nation's cultural traditions is tantamount to rejecting the traditions themselves. It signals the end of a lean, hardworking, straight-shooting American ideal, and its replacement with a new ideal. That new ideal still involves freedom, but the rejection of the paradigm embodied in a respect for nature ushers in a new sense of freedom appropriate to a new age. That age is symbolized by powerful cars, airplanes, and fast food establishments; it represents a type of freedom identified with ease of life, dependence upon gadgets to give more leisure—a freedom to do what one wants at any time with minimal effort.

Second, the argument assumes that American cultural history and the values embodied in it are of great value. Americans are better off, in some important sense, because those values are a

[25] Sagoff, "On Preserving the Natural Environment," pp. 228-229.

part of their traditions. They would be discarding something of great value if their paradigms of freedom, honesty, and self-reliance were lost. This is the overwhelmingly important contribution made by Sagoff's essay, embodied in the first feature of his account: he recognizes that metaphorical experience functions not merely as a *fulfillment* of fixed desires but also as a *shaper* of those very desires. Experiences of nature have value not merely for consumption; they are not merely the stuff from which preferences are fulfilled. They are, rather, the stuff of which ideals are made, and the preferences that are dominant in a culture are determined by its ideals.

What Sagoff establishes is that the process is a spiralling one. Destruction of the natural heritage central to American culture eliminates an aesthetic element essential in the formation and perpetuation of the ideals which gave that culture its unique, new-world distinctiveness. These ideals, when writ large across a culture, determine the sorts of options available to its members. If these ideals lose currency, it can be expected that less will be done to protect the wild places and wild species that provide the occasion for the aesthetic experiences, the grasping of the metaphors, that are essential to the formation of the ideals. American values will shift toward an ideal that is less shaped by wild and unspoiled nature and, as the ideal changes, preferences will change. There will be less support for protecting unspoiled nature. And so on and on the process will go.

For those, therefore, who value the traditional ideals of American culture there is a powerful argument for protecting wild species and wild things. That argument transcends questions of preference *satisfaction* and derives from the preference *formation* role played by experiences of wilderness. It has to do with the sort of culture we wish to create and concerns the quality of life toward which we and our posterity will aspire.

10.5 Cultural Relativism

Sagoff's aesthetic argument for preserving species as embodiments of traditional American values is very powerful politically, at least in circles where those values are esteemed.[26] But Sagoff's form of argument puts tremendous pressure on those values: How are they, in turn, to be justified? Here Sagoff's position becomes untenable. What is to be said to the worshippers of high-tech culture? They are likely to admit, even exult in, the suggestion that they are harbingers of a new cultural age. They are also likely to take themselves to be making a positive contribution to American culture. They will ridicule those who mourn the loss of traditional values as hopeless conservatives. If the causal connection between the destruction of wilderness and the destruction of traditional values is taken seriously, they might have reason to argue in favor of bringing out the bulldozers.

Values cannot be justified merely because they are traditional. That, in itself, is no argument for them. The assumption that traditional values should be supported over new and modern values seems just a conservative, perhaps even elitist, reaction.[27] And even if one could argue that Americans should continue to support traditional, nature-oriented values, an aesthetic argument so based provides no reason for other cultures to do so.[28] Other cultures seem to do fairly well with a different set of values. Thus, while Sagoff's argument can provide reasons for Americans to preserve species, no comparable arguments are available to persuade citizens of other nations to preserve species within their national boundaries. Indeed, conflicts in values seem inevitable, compounded by problems of sovereignty: What is to be done when the values of one culture favor preservation of a species whose range is in another country?

These are serious problems for Sagoff's argument. Its narrowness limits its usefulness for species preservationists because they want to generalize their appeal to all countries, to all cultures, and to all individuals. A relativistic argument such as Sagoff's can be persuasive ultimately, in the broadest context, only if overwhelm-

---

[26] Tom Regan, "The Nature and Possibility of an Environmental Ethic," *Environmental Ethics* 3 (1981): 29.
[27] Many writers have similarly criticized Sagoff. See, for example, ibid., p. 30.
[28] Ibid.

ing arguments support the set of values in question. These Sagoff has not given.

It is crucial to note, however, that the first premise of Sagoff's argument, the recognition of a connection between changing values and changing cultural symbols, need not be rejected along with the second. This premise can stand as a basis for a more general argument, provided the values served by such symbols can be made more universal and less dependent upon American cultural history. Such a strategy implies accepting Sagoff's first premise, that ideals are shaped by aesthetic, metaphorical experiences, and then searching for ideals that are shared by all human cultures. If those common traditions and ideals are also shaped, clarified, and supported by aesthetic encounters with wild species and wild lands, then there will exist an argument that, like Sagoff's, recognizes the importance of experience of nature in shaping values but that, unlike his, relies on general human values important to all cultures.

10.6 *Nature and Common Values*

When environmentalists in general and species preservationists in particular criticize the overmaterialistic and overconsumptive nature of contemporary values, they do not intend merely to say that those values are inconsistent with one tradition in American intellectual history. Nor did the founders of that tradition see themselves as creating a subjective set of values incapable of general application. While Thoreau and, especially, Emerson emphasize the new possibilities for a distinctive culture to emerge in the new world, they leave no doubt that such a culture would attain an objective, universal truth:

> We must trust the perfection of the creation so far as we believe that whatever curiosity the order of things has awakened in our minds, the order of things can satisfy. Every man's condition is a solution in hieroglyphic to those inquiries he would put. He acts it as life, before he apprehends it as truth. In like manner, nature is already, in its forms and tend-

encies, describing its own design. . . . Let us inquire, to what end is nature?

All science has one aim, namely, to find a theory of nature. . . . To a sound judgment, the most abstract truth is the most practical. Whenever a true theory appears, it will be its own evidence. Its test is, that it will explain all phenomena.[29]

These are not the words of a relativist. Emerson is extolling the virtues of a fresh start in the new world not because he seeks a distinctive philosophy valid only in a new context but because he believes the old ways have concealed the path to a complete, consistent, and correct view of nature. What is to be abandoned is not the old objective of universal truth but the old way toward that objective, littered with the encrustations of worn-out cultures. The new start is noteworthy because truths imperceptible in the corrupt societies of the old world lie open to the understanding of unspoiled humans more in touch with nature's harmonies.

When environmentalists represent themselves as the harbingers of a new world view and a new set of values, they may intend to create a new path but toward a universal and objective destination. They believe that it would be a good thing—an unqualifiedly and nonrelativistically good thing—if the human species were to adopt a world view and set of values that placed them in harmony with the workings of nature. To the extent that our current culture has failed to achieve such harmony, they believe it has taken a wrong road and not just a road wrong "for us"—an objectively wrong road.

They believe that the values this culture now seeks are inconsistent with a rational and realistic world view. If humans would see the world aright, if they would see themselves and their activities in proper perspective, they would reject current felt preferences and strive toward the adoption of higher, less materialistic, and less consumptive values—a set of considered preferences appropriate to a modern, rationally defensible world view.

Now, critics have every right to be skeptical here. My argument

[29] Emerson, "Nature," pp. 35-36.

so far has been generated from, first, listening to what species preservationists are saying and, second, developing concepts and principles that could use to support those statements. It is a far more difficult task to support such a framework, a task that cannot fully be undertaken here. In the remainder of this chapter, I will show, in a preliminary manner, how species preservationists might make their case. My goal will be to show which premises are essential and how they might be made plausible rather than to produce knock-down arguments. I offer a blueprint rather than a finished building.

The conceptual apparatus outlined in this book, the distinction between felt and considered preferences and the attendant disjunction of demand and transformative values, provides the scaffolding for building the edifice. But it provides no firm foundation. For that, substantive premises concerning the nature of what is real and what is valuable must be supported. Only given these can the critique of felt preferences and the promulgation of considered ones proceed.

First, species preservationists, and environmentalists more generally, believe that the human species has evolved as other species have, within complex and interrelated ecosystems. While modern technology may often obscure this fact, *Homo sapiens* remains dependent upon biotic and abiotic environments in countless ways. Our present needs and characteristics and the very evolutionary necessities embodied in our genes are the result of past adaptations to environmental conditions. Further, the future possibilities among which we will choose are likewise determined by the interaction of current genes and future environmental conditions. Two great ideas come together here: Darwinian biology has taught us that humans are, basically, evolved animals; ecology has taught that evolution works within almost unbelievably complex and interrelated organic systems on interlocking levels ranging through molecules, cells, organs, organisms, habitats, ecological systems, the biota as a whole and, ultimately, the abiotic system. All levels of life interact within the parameters set by the abiotic conditions that at once determine and are determined by them. In a sense these two great ideas, once combined,

204

provide an ontology for the ecological world view, a recognition of certain things as most basic and determinative.

If the first principle of the ecological world view sets down an ontology, the second principle provides an epistemology of sorts. As was just noted, preservationists believe that the natural world is extraordinarily complex. The levels of interrelationships are multifarious, and minute alterations in the smallest elements making up a system initiate further alterations cascading throughout the system, and these affect again the elements themselves. An epistemology appropriate to this belief is cautious and skeptical in its particular claims to knowledge and in its assessment of knowledge generally. The more we understand, the more we realize how little we understand. Where every fact, no matter how specific, causally depends upon alterations in a system that it in turn alters, dogmatism is not a virtue. The ecological world view and its attendant skepticism represent the final undermining of Cartesian epistemology. The search for unquestioned and unquestionable premises, for "first truths," for certainty, is at an end. Theories, however grand, become models to help us to understand how change courses through systems. First principles are heuristic starting points, as open to alteration as are the effects they hypothetically predict. While one need not doubt that change takes place according to some coherent principles, the principles are best seen as means to understanding reality rather than as constituting it. From quantum mechanics to ecological theory the epistemological lesson is the same: each action, even if it is a measuring action, changes the system in which it intervenes. Human knowledge alters the nature of the systems known, and this fact follows (at least loosely) from the ontology of the preservationist: human knowledge occurs within and affects the system of nature.

Third, preservationists believe that the ecological world view, embodying these ontological and epistemological principles, lends credence to a generally ecological approach to values and objectives.[30] The approach is one of humility. The variety and in-

[30] I am here intentionally vague regarding the type of inference employed. The pitfalls of suggesting that values follow from nonevaluative premises alone are

tricacy of nature are so far beyond our current comprehension that, while more knowledge usually helps, so much more is needed that wisdom implies the pursuit of cautious goals. Action based upon knowledge of a single relationship abstracted from a complexly interrelated system is dangerous. Manipulation of single variables produces unforeseen results. This is what environmentalists mean when they caution that "you can never do just one thing." But humility need not imply despair. If humans are highly evolved animals, then the species has succeeded, through the ages, in developing a viable relationship with its biotic and abiotic environment. The natural history of *Homo sapiens*, viewed as a highly evolved and highly intelligent but physically dependent being that has survived in a hostile world, can stand as a guide to human behavior.

Does this philosophy imply quiescence, a paralysis of human will in the face of looming human problems? No. The human species would never have survived had it not used its intelligence to alter the world in which it lived. Even lacking complete understanding of the components and interrelationships of the system on which we depend, there are bases for determining wise human action. Nature's system works because of harmonies and balances within the system. Natural systems are geared for change by component parts: systems are dynamic, not static. But they are more geared for some types of change than others. Incremental changes that mimic natural processes, such as taking older and weaker organisms to fulfill human needs, or controlling an outbreak of a pest species by encouraging its natural enemies, work within the established patterns of the system of nature. These changes set off compensating mechanisms already in place; information flows through the system, allowing other organisms to react.

---

well known. Environmentalists appear to believe that there is a sense of "fit," of appropriateness, between ecological science, philosophical principles based upon ecology, and certain moral principles implying the protection of nature. If they merely claim that science and nonevaluative philosophical principles give rise to certain ethical principles and make them more plausible, they need not commit the naturalistic fallacy.

Abrupt changes, changes with no parallels in normal biological or climatological processes, scramble the feedback messages and leave the system in disarray.[31] Ecosystems have few mechanisms for adapting to expanses of concrete, and these act too slowly to protect the system paved over from simplification and collapse. A bulldozer's clearing of a small patch can be remedied in a few seasons; it is similar enough to the clearing caused by the fall of a large tree, which is subsequently filled through processes of colonization and succession. But if a whole area is defoliated, even the colonizers are beaten back from their frontiers. The process of recovery is slow and incomplete; opportunists take and hold possession for a longer time, making it more difficult for full regeneration to take place.

Thus, the ontology and epistemology of the ecological world view give rise to a positive value—that of harmony with nature and nature's way. It is good, in this view, to do things in a way that mimics nature's patterns; it is good to promote the natural processes that, if not interrupted, produce greater diversity; it is good to introduce alterations slowly enough to allow nature to react. And it is bad to thwart those natural processes, to interrupt well-established patterns, to introduce irreversible changes.

10.7 Response to a Skeptical Objection

Besides objecting that these are vague platitudes (an objection I hope to answer with a detailed example in the next chapter), the skeptic may argue that this system only replaces short-term selfish motives with a more enlightened sense of self-interest. In one sense this skepticism is justified. Transformative values are human (anthropocentric) values. These values are not limited by intrinsic values discovered in, or attributed to, nonhuman nature. But the distinction between felt preferences and the considered preferences that replace them leads to an important qualification of the skeptic's point.

[31] See Eugene Odum, *Fundamentals of Ecology*, 3d ed. (Philadelphia: W. B. Saunders Company, 1971), p. 269.

A considered preference is one that survives after a complex process of analysis and self-criticism. Such a process might take a number of forms, but it will be useful to sketch one of these. For concreteness let us consider Jane, a young woman who lives near home and helps her parents with the family business. Her parents are generous, and she is able to afford many luxuries, including expensive clothes, a comfortable apartment, and a sports car. But she is vaguely dissatisfied.

One evening, more from boredom than any other motive, Jane accompanies a friend to a meeting of a conservation group. Jane has always liked birds and is inspired by slides of whooping cranes. It is disturbing to her that human activities have brought this species to the brink of extinction, and she is angered by a government plan to further compromise its remaining habitat. Jane makes a considerable donation to the conservation group and hardly misses the amenities she might have purchased instead. Soon she becomes an enthusiastic volunteer and spends all of her free time in conservationist activities.

Finally Jane decides to attend a distant university that has a special program in wildlife protection. To her surprise Jane's parents adamantly oppose the plan and refuse both emotional and financial support. When they learn that she is still considering the move, her father asks, "You mean a few birds mean more to you than your own parents?"

Faced with that question, Jane begins some real soul-searching. Does she value endangered wildlife more than she values her parents? She remembers that one speaker at the conservation meetings said that all living things have intrinsic value. She had been puzzled and had asked what that meant. He had answered that all living things have the same kind of value that humans do. That didn't help much, and it certainly doesn't convince her that birds are more valuable than her parents.

Jane has reached an impasse. On the one side are her parents and the comfortable life style that working for them guarantees. On the other are her newly developed ideals to protect wild animals and the environment. Eventually she realizes that she will never resolve the issue framed in terms of whether the animals she hopes to protect have sufficient intrinsic value to counterbalance her love for her parents.

Then she begins to see the question in quite different terms. She realizes that, since her first attendance at the conservation meetings, *she* has changed. She has begun studying ecology texts and now finds a walk in the woods stimulating and satisfying. She also notices that she feels more comfortable in her "Save the Animals" sweatshirt than in an expensive new dress. A day of shopping, once her favorite activity, leaves her less satisfied than a day working as a volunteer for a conservation cause. Without really thinking about it, she has realized that there are more important things than material possessions.

In the end she concludes that, however much she loves her parents, the decision is hers. And, while the discussions she's had about intrinsic value in nature do not help her to make her decision, she knows that her ideals have changed. The process began when she saw the slides of whooping cranes. It continued when she read ecology texts that talked about the interdependencies of all living things, and when she did volunteer work. At the same time her tastes and preferences changed. And she believes that she is a better person for these changes.

Jane decides to sell her sports car to cover the first year's tuition. With a small scholarship, a small loan, and a part-time job, she is able to cover her surprisingly modest material needs, and she eventually graduates. While her parents remain incredulous and often remark that she'd be welcome to return to the business, Jane knows she won't accept their offer.

It is important not to suggest that every person faces exactly the same self-examination in exactly the same way, but Jane's example allows us to focus on some key aspects of the processes of considering preferences. First, the process often begins with a new experience, or with a new orientation toward a common experience. Attending the conservation meeting and seeing slides of whooping cranes planted a seed that grew and flowered in Jane's consciousness. Second, Jane began to alter her behavior by donating money and by volunteering to work for preservation goals before any radical reexamination took place. Some people may begin by reconsidering their deepest values and principles, but Jane's concrete approach is more common. Third, when Jane discovered that her shifting interests brought new meaning to her life, she began to feel uncomfortable with her previous life goals and the satisfactions she had sought. Finally, Jane came to the realization that the new goals and ideals that had taken shape were objectively better than the old ones. It was not crucial that Jane determine that wild animals have intrinsic value—it was only necessary that she conclude that her new values, which avoid unnecessary consumption, are better than her old, materialistic ones. She is especially happy to realize that pursuing her new values also protects species, both because she no longer wears a fur coat and because her new career will make a positive contribution to the cause of species preservation.

Species preservationists view their activities similarly. If they believe that the ecological world view represents a more accurate picture of the world and that the value system suggested by it is objectively better than the value system of materialism and conspicuous consumption, then they will value endangered species and natural ecosystems for their role in transforming human world views and human value systems. Jane's story began with an encounter with an endangered species. If the story of her individual transformation is plausible, species preservationists can hope that a parallel transformation might occur in our entire society. But it will not occur if nature is so altered that encounters with wild species become unlikely. Species preservationists should emphasize the value of wild species, especially endangered ones, as catalysts for the reconsideration of currently consumptive felt preferences.

As long as species preservationists agree that a nonmaterialistic value system is objectively better than one based upon unlimited consumption, it does not matter whether they also agree that nature has intrinsic value. The transformative value of wild species and natural ecosystems provides adequate reason to preserve them, quite independent of whether nature has intrinsic value.

In one sense, then, the skeptic is correct. Jane rejects her consumptive life style for self-oriented reasons. She does not rely on attributions of intrinsic value to nonhuman species to force a change in herself. Nor does she sacrifice her interests for the competing interests of other species. Has she, then, merely substituted a somewhat enlightened self-interest for an unenlightened self-interest? By a complex process impossible to separate into sharply defined stages, Jane's preferences, if spread throughout the entire population, would no doubt improve the long-term survivability of the species. But this realization was not determinative in Jane's story, nor is this realization the "justification" for the changes in her attitudes. Jane concluded, after her process of change, that her new values were objectively better than the consumptive ones she gave up, meaning that she had a better and richer life as a result of the transformation. The value of the change is measured in human terms, but that conclusion is merely a restatement of the original premise, that transformative values are anthropocentric.

Environmentalists accept the "ecological world view" as sketched in Section 10.5. They also believe that encounters with wild species can precipitate changes in human consciousness, alterations in world views sufficient to create a new ontology, a new epistemology, and a new approach to value. If they also believe that the new, less materialistic values that are thereby created are objectively better than the materialistic, consumptive values they replace, they should value all wild species, including endangered ones, for their transformative value. On this anthropocentric basis they can argue that species should be preserved, regardless of whether they also believe that species have intrinsic value.

This, then, is a coherent and complete argument for protecting species. It recognizes the demand value that species have, while also insisting that experience of wild species can enlighten demand values by initiating a process of reexamination. Such experiences interact with, and support, ideals that undermine a materialistic, consumptive style of life. Attributions of intrinsic value to nonhuman species might be included in those new ideals but need not be.

Species preservationists can argue that their world view—its ontology, its epistemology, and its value of harmony with nature—is a rational response to the world as it is encountered one hundred years after Darwin's formative discoveries. In this way they can offer a framework within which felt preferences can be criticized and demand values transformed by a process that shares essential features with the alterations in Jane's consciousness. Species preservationists can argue that overly consumptive demand values are less rational because their fulfillment threatens the system within which the human species has evolved and must continue to exist.

But the caution against destroying that system, the advocacy of harmony as a better course than hubris, is not understandable solely in terms of demand values, even when the demand is for survival. The risks to survival taken by the human species when it flouts natural constraints are only symptomatic of a deeper crisis, according to those who fully accept the ecological world view. They are symptomatic of a rejection of a deep truth about ourselves—that we differ from other living things only in the nature of our adaptations. To destroy other species and the ecosystems in which they evolve is to treat our own past with contempt. It is to forget who we are. If we do not know who we are, it is unlikely that we will adopt a rationally justifiable value system.

According to the ecological world view, then, the human species faces constraints imposed by its dependence upon natural ecosystems. Some demand values are consistent with a recognition of this fact and others are not. Those that are not must be subjected to rational criticism. But the sort of ecological world view within which these values can be criticized cannot be articulated a priori. It can only be worked out by deepening our understanding of ecological relationships—how species function in natural ecosystems. If species are destroyed and ecosystems degraded, the human race cannot gain the knowledge necessary to formulate the details of such a world view. Wild species and pristine ecosystems teach us about ecological relationships and provide analogies and metaphors that give us self-knowledge. They also provide the occasions for forming and criticizing our values, as felt preferences are measured against the evolving world view. In this way they have transformative value.

Sagoff links the mechanisms by which these processes take

212

place to special features of American culture. But problems of how to exist in a limited world, of how to understand that world, and what value to place on it are not uniquely American problems. If the theory of evolution is correct in its broad outlines, it would not be surprising if significant guidance in solving these problems lies in a deep understanding of the roots of the human species. Other species provide the best key to understanding our evolutionary history, and experiences of them can transform human value systems. Transformative values thereby afford a considerable argument for preserving other species.

E L E V E N

A COHERENT RATIONALE FOR
SPECIES PRESERVATION

11.1 Resources versus Nonresources

Species preservationists often argue or assume that reasons for
protecting species can be grouped into two basic categories, designated variously as economic versus noneconomic, utilitarian
versus nonutilitarian, prudential versus ethical. These various
distinctions, while related, need not delimit exactly the same categories, and they all suffer from certain ambiguities. But references to some such bifurcation persist.

My purpose in this chapter is to integrate the diverse reasons
discussed in preceding chapters into a coherent rationale for preserving species. It would be useful if one element of this integra-
tion were a typology of general rules for grouping, organizing,
and perhaps even setting priorities among the goals cited in the
disparate rationales offered. David Ehrenfeld has suggested for
this purpose a distinction between "resources" and "nonresources," and it will be useful to begin by examining his dichotomy.[1]

Ehrenfeld begins by decrying the overuse of the term "resource" and suggests that it be defined "very narrowly" as "reserves of commodities that have appreciable money value to people, either directly or indirectly."[2] In this narrow conception of
resources, which Ehrenfeld favors, a species, viewed as an ongo-

[1] David W. Ehrenfeld, "The Conservation of Non-Resources," *American Scientist* 64 (1976): 648-656, and *The Arrogance of Humanism* (New York: Oxford
Univeristy Press, 1981).
[2] Ehrenfeld, *Arrogance of Humanism*, p. 178.

ing, changing collection of individuals, is *not a resource, by definition.*

Ehrenfeld's distinction has clear relevance to the distinction between economic and noneconomic reasons. He sometimes uses the phrase "economic and selfish reasons" interchangeably with "resource" reasons, trying to illuminate the distinction between prudential and ethical reasons,[3] and he uses the resource/nonresource dichotomy in tandem with references to "the humanistic system," a set of assumptions about the predominance of human power and value.

Ehrenfeld is concerned that environmentalists place too much emphasis on resource arguments, asserting that "there is no true protection for nature within the humanistic system—the very idea is a contradiction in terms."[4] His arguments for this conclusion can be grouped under two headings: (1) reliance on some resource arguments undermines environmentalists' objectives, and (2) exclusive reliance on resource arguments encourages distortions and rationalizations that weaken environmentalists' credibility.

Arguments clustering around (1) include: (a) an objection to "piecemeal conservation of things in Nature";[5] (b) concern that resource arguments lose force as new inventions provide artificial substitutes for goods and services once derived from nature;[6] (c) fear that resource arguments encourage rankings and choices that are "likely to set Nature against Nature in an unacceptable and totally unnecessary way";[7] and, finally, and most basically, (d) that once one treats a part of nature as a resource, it has value within the humanistic system and is subject to overexploitation. Applying this argument to attempts at assigning economic value to salt marshes, Ehrenfeld says: "But discovering resource value can be dangerous; in effect one surrenders all right to reject the

[3] Ibid., pp. 208, 210.
[4] Ibid., p. 202.
[5] Ibid., pp. 189, 192.
[6] Ibid., p. 201.
[7] Ibid., pp. 203-204.

humanist assumptions."[8] Ehrenfeld is also concerned about the credibility of environmentalists, citing the claim that nature is so highly interconnected that no one part can be lost and the diversity-stability hypothesis as cases where environmentalists have distorted facts to support their position, only to be found out and have their future credibility undermined.[9]

While a number of Ehrenfeld's arguments (especially those in the first group) are persuasive, I find his use of the central dichotomy between resources and nonresources puzzling at best. His concept of resources refers to reserves of commodities, that is, categorizes certain objects as having money value. According to this conception, the existing blue whales and the standing crop of redwoods are resources, but the species are not. Likewise, the actual outputs of an ecosystem such as blueberries from a blueberry bog would be a resource, but the ecosystem would not. Ehrenfeld's narrow concept of resources partitions nature into useful natural *products* and their *sources*, the species and processes that yield them.

This approach is unique in two respects. First, most writers who have struggled to capture the generic dichotomy here at issue distinguish two types of *reasons* for preserving natural objects and ecosystems. It is then assumed that some objects will be susceptible of protection by appeal to one type of reason, some by appeal to the other, and some by both. But Ehrenfeld introduces his dichotomy as distinguishing two types of *entities*. It separates natural products from the sources of those products, species and ecosystems. It is a mistake, given his definitions, to refer to species and ecosystems as resources at all.

Second, using his narrow definition to exclude reference to species and ecosystems as resources, he seems at first to deny that they can even be defended as being useful to humans, hence his claim that protection for nature within the humanistic system is a contradiction in terms. He concludes that nonresources such as

---

[8] Ibid., pp. 199, 202. Also see my discussion of this subject in Section 3.2.
[9] Ibid., pp. 190, 193.

species and ecosystems must be defended on the grounds that they have intrinsic value: *"they should be conserved because they exist and because this existence is itself but the present expression of a continuing historical process of immense antiquity and majesty.* Longstanding existence in nature is deemed to carry with it the unimpeachable right to continued existence."[10]

This position is coherent. Species and ecosystems are not resources; their value exists outside the humanistic system. Since they all have intrinsic value, they avoid "piecemeal" conservation, they cannot lose value as substitutes for them are invented, they cannot be set against each other or ranked, and they are not subject to overexploitation. The dangers cited in arguments (1a) to (1d) are avoided.

But implementing this suggestion would be very costly for species preservationists. It entails the abandonment of all arguments based on the usefulness of species and ecosystems to humans, such as those developed in Part A. The unlikelihood of such wholesale abandonment is illustrated by Ehrenfeld's own inability to adhere consistently to his conclusion:

I [do not] reject resource arguments when they are valid. The Amazonian rain forest, the green turtle, and many other forms of life are indeed resources; they contribute heavily to the maintenance of human well-being. The prospect of their loss is frightening to anyone with ecological knowledge, and it is not my aim to make it appear less so. But this is only one of the reasons for conservation, and it should not be applied carelessly, if only because of the likelihood of undermining its own effectiveness.[11]

Here Ehrenfeld has violated his own definitions by treating ecosystems and species as resources. Further, he has violated the spirit of his conclusion that species and ecosystems have a right to existence because he adopted that conclusion precisely in order to

[10] Ibid., pp. 207-208 (Ehrenfeld's emphasis).
[11] Ibid., p. 200.

avoid treating them as resources. And the reason for avoiding this was that to treat them as resources is to place them within the humanistic framework of value where they would be subject to the threats outlined in arguments (1a) to (1d). The logic of these arguments admits of no degrees—they are designed to show that once species or ecosystems are assigned *any* value within the humanistic system, they are vulnerable to threats resulting from the inconstancy and exploitative nature of that value. Only if these arguments admit of no degrees could Ehrenfeld claim that protection of nature within the humanistic system is a contradiction in terms.

Ehrenfeld thus holds two contradictory positions. On the one hand, he defines resources narrowly as including only standing stocks and products. Species and ecosystems are nonresources, to be defended on the basis of their intrinsic value. Because he is unwilling to abandon all appeals to the usefulness of species to humans, however, he also suggests that one can appeal to resource justifications for preserving species as long as they are valid. The contradiction appears less blatant because, although his central dichotomy is between two types of things (a partitioning of nature into resources and nonresources), he also lapses into use of the more usual distinction between two types of justifications (resource justifications and nonresource justifications) for preserving species and ecosystems. But he cannot have it both ways. According to his narrow definition, either species and ecosystems are resources or they are not. If they are, then he has given them value within the humanistic system and they would appear to be susceptible to the ravages he describes as characteristic of that system. If they are not, then he must entirely give up resource arguments for their preservation.

The incoherence of Ehrenfeld's position results from two common confusions acting in tandem. First, he seems to believe that one must choose between attributing intrinsic and instrumental value to an object, that no object can be valued for its intrinsic value and simultaneously for its usefulness. Hence, the argument just quoted: "But discovering resource value can be dangerous; in effect one surrenders all right to reject the humanist assumptions."[12] This is surely wrong. One can assign instrumental value to an object without automatically denying that it has value beyond that usefulness. If this were not so, I could not value my garage mechanic for his reliability, honesty, and skill in fixing my automobile without thereby denying him respect as an independent locus of fundamental value. Attributing intrinsic value to an object limits the *ways* in which that object can be used but need not prohibit all use of it.

Second, Ehrenfeld is convinced that one must either attribute intrinsic value to an object or else leave it without any protection from the vagaries of human consumptive demands. This belief leads him to conclude that protection by a humanistic system is a contradiction in terms. This too is a confusion. If there were widely accepted and well-supported principles and rules governing the behavior of humans in regard to nature, such as those related to the ideal of living in harmony with the biotic world, these could distinguish acceptable and unacceptable uses of nature. One need not attribute intrinsic value to a neighbor's property in order to have good reason not to destroy it. Nor need one attribute intrinsic value to nature in order to have good reason not to use it destructively.

These two confusions are common in writing about species preservation. Taken singly or together, they encourage the view that environmentalists face a serious dilemma: either nature has intrinsic value capable of implying severe restrictions on human use of it or there are no restrictions (except those implied by fairness to other humans) on human treatment of nature at all. But this dilemma, however influential, is a false one.

Its influence can be seen in Tom Regan's rejection of an initially attractive, but anthropocentric, position on his way to defending the view that any adequate environmental ethic must attribute intrinsic value to nature. Regan introduces an argument—"the offense against an ideal argument"—that he takes to be characteristic of such an anthropocentric position: "The argument alleges, quite apart from how those who treat nature end up treating

[12] Ibid., p. 201.

219

other humans, that those persons who plunder the environment violate an ideal of human conduct, that ideal being not to destroy anything unthinkingly or gratuitously."[13]

Regan criticizes this view as follows:

> If we are told that treating the environment in certain ways offends against an ideal of human conduct we are not being given a position that is an alternative to, or inconsistent with, the view that nonconscious objects have a value of their own. The fatal objection which the offense against an ideal argument encounters, is that, rather than offering an alternative to the view that some nonconscious objects have inherent value, it presupposes that they do.[14]

Regan states, prior to this conclusion, three propositions that are intended to support it. (1) The fitting way to act in regard to X clearly involves a commitment to regard X as having value. (2) An ideal that enjoins us not to act toward X in a certain way which denies that X has any value is either unintelligible or pointless. (3) Ideals, in short, involve the recognition of the value of *that toward which* one acts.[15]

But these assertions are all ambiguous, of course. Does "value" mean "intrinsic value" or "instrumental value"? Claims (1) through (3) would be unexceptionable if they mean that ideals regarding the treatment of an object presuppose that the object has intrinsic *or* instrumental value. But surely Regan cannot intend to draw the strong conclusion that he does from these premises so interpreted. So, proposition (1), for example, must be interpreted as claiming that if someone believes that there is a morally fitting way to act in regard to some object X, then that person accepts a commitment to regard X as having *intrinsic* value.

According to this interpretation, (1) to (3) are all false. I can accept that there is a morally fitting way to act in regard to my neighbor's plate glass window, refraining from throwing a brick

---

[13] Tom Regan, "The Nature and Possibility of an Environmental Ethic," *Environmental Ethics* 3 (1981): 25.

[14] Ibid., pp. 25-26. Regan uses "inherent" value in place of my "intrinsic" value. I believe that this substitution does not affect the argument here.

[15] Ibid., p. 25.

through it, without thereby accepting any commitment to accord intrinsic value to the plate glass window. I am not, contrary to Regan's (2), thereby committed to anything either unintelligible or pointless. And I am not, contrary to his (3), recognizing the intrinsic value of the window. One can institute proscriptions against behaviors regarding objects without implying that those objects have intrinsic value.

There are moral reasons against throwing a brick through my neighbor's window. In this example the reasons derive from my neighbor's intrinsic value and her preferences, augmented by the convention of private property. But not all human-oriented moral reasons are based on individual preferences in this way. Nature has transformative, as well as demand, value. If I believe that the child who once enjoyed destroying birds' nests is a better person for coming to appreciate living birds, I value wild birds because they provide *occasions* for the consideration and transformation of felt preferences.[16] And, while I might believe that the child's value system is improved because he came to recognize the intrinsic value of nonhuman species, I may also believe that his altered value system is better for anthropocentric reasons. For example, I might believe that humans who protect rather than destroy other living things are less likely to be violent in their dealings with other humans. Or I might believe that hobbies such as ornithology make people less materialistic and less oriented toward consumption. If I believe that violence and materialism are morally inferior to nonviolence and contemplation, I should therefore value the wild birds as providing occasions for the uplifting of human attitudes and values. In many different ways, then, encounters with wild species contribute to the formation and reformation of our attitudes and values—and, in turn, our preferences.

11.2 Consumer Preferences and Citizen Commitments

When Regan and Ehrenfeld envision only two possibilities, one that other species be accorded intrinsic value and the other that they must receive inadequate protection from human misuse,

16 See Section 10.2.

221

their reasoning is akin to that of benefit-cost analysts. As I have defined the BCA approach, it rests on the assumption that all relevant values can be assigned some nonarbitrary quantitative value. This quantitative assumption presupposes a unitary scale for measuring values—that of individual preference satisfactions. Dollar figures are operational measures of these preferences. When some environmentalists recoil from anthropocentrism, adopting a position they admit they do not understand, one suspects that their real target may be the assumption that all values must be expressed quantitatively.

These environmentalists fear that nature cannot be preserved without appeal to qualitative values capable of overriding actual and quantifiable individual felt preferences. And since they implicitly accept the BCA assumption that all human values must be perceived as actual preference satisfactions, they see no way to invoke qualitatively expressed values without denying anthropocentrism altogether. But they could as well reject the assumption that all human values are expressible in terms of quantifiable measures of individual preference satisfactions. A narrow view of human values, in other words, encourages environmentalists to look to nonhuman sources of value to justify their preservationist policies. A contraction of human values to measurable and quantifiable preference satisfactions—a methodological necessity for BC analysts—also encourages environmentalists to embrace nonanthropocentrism. Presently I will argue that human transformative values cannot be expressed in terms of quantitative preference satisfactions. If this argument is persuasive and if environmentalists appreciate the importance of transformative values, they can reject the BC analysts' assumption by appealing to human, nonquantifiable values. The dilemma troubling Regan and Ehrenfeld is thus avoided, and an important motivation for embracing nonanthropocentrism dissolves.[17]

[17] I am not suggesting that there are no other motivations for nonanthropocentrism. Arguments for nonanthropocentrism that note positive characteristics of nonhuman species and argue that these justify attributions of intrinsic value are not affected by this analysis. See, for example, Holmes Rolston, III, "Are Values in Nature Subjective or Objective?" *Environmental Ethics* 4 (1982): 125-151.

In Part A, I showed how BC analysts confront values that are not readily expressible in dollar figures. They acknowledge that some values have no market.[18] Consequently, they correct such "market failures" by establishing hypothetical prices designed to represent what individuals would be willing to pay for the protection of those values (WTP or WTA), if there were a market for their protection.[19] I proceeded to argue that the WTP and WTA criteria cannot reflect adequately the complex and poorly understood benefits humans derive from wild species and natural ecosystems. Quantified values applied to species are, therefore, unreliable expressions even of human demand values.

I have now established that wild species and natural ecosystems have transformative as well as demand value for humans. Can these transformative values be given adequate expression in BCAs? If they cannot, this failure will constitute another reason to reject the assumptions essential to the BCA approach to valuing species (as I have defined it).

The assumption that all human value can be expressed in quantitative terms and its attendant presupposition that all human values can be expressed as human preference satisfactions encourage a picture of human desires as sovereign.[20] Whatever humans want is, by definition, valuable. It is difficult to see how limits on the appropriate pursuit of preferences can be imposed within such a system. Hence, Ehrenfeld's concern that treating species

---

The current analysis does apply, however, to arguments relying only on indirect analogies. See Section 8.1.

[18] See A. Myrick Freeman, III, Robert H. Haveman, and Allen V. Kneese, *The Economics of Environmental Policy* (New York: John Wiley and Sons, Inc., 1973), p. 65. Also see Section 2.2 above.

[19] See Lowdon Wingo, "Objective, Subjective, and Collective Dimensions of the Quality of Life," in *Public Economics and the Quality of Life*, ed. Lowdon Wingo and Alan Evans (Baltimore: Johns Hopkins University Press, 1977), pp. 13-14.

[20] See, for example, William F. Baxter, *People or Penguins: The Case for Optimal Pollution* (New York: Columbia University Press, 1974); and Larry Ruff, "The Economic Common Sense of Pollution," in *Microeconomics: Selected Readings*, 4th ed., ed. Edwin Mansfield (New York: W.W. Norton and Co., Inc., 1982). On the sovereignty of human preferences, see, for example, Dennis Mueller's account of Kenneth Arrow's rule of the "unrestricted domain," as presented in *Public Choice* (Cambridge: At the University Press, 1979), p. 194.

and ecosystems as resources, as means to fulfill human preferences, leaves them with no firm protection. If human value is uni-
dimensional and consumptive, little room remains for principles
and rules limiting consumption.

Of course the BCA approach has some decided advantages,
both practical and theoretical. Practically, this approach allows
the measurement of human values on a single, quantifiable (usually monetary) scale, thereby presenting data in a more easily usable form. Theoretically, it allows policy makers to avoid entan-
gling normative arguments: policy can be designed to implement
what individuals say they want. In spite of technical problems involved in placing monetary figures on some values,[21] advocates of
the BCA model believe that the price a rational actor is willing to
pay to obtain or protect an object can stand as a measure of its
value. To the extent that willingness to pay is a good indicator of
human values, policy analysts can claim not to judge, but only to
report, the preferences that individuals express. Consequently it
is possible to claim that the analysis is "value neutral."

While these practical and theoretical advantages cannot be denied, many commentators have noted that the model is problematic as a complete understanding of values as they affect policy
because the unitary scale ignores important differences among
human values.[22] For example, consider the value individuals
place on their freedom to participate in the political process, the
freedom to vote in free elections. According to the assumptions
underlying the BCA approach, there must be some dollar amount
that a person would accept in exchange for his right to participate. It seems rational, however, for an individual to refuse to
place a dollar figure on a basic right of this sort because the right
to participate determines the ability to protect future interests
and is therefore valued on a different scale.

[21] See above, chapters 2 to 4.
[22] See Mark Sagoff, "At the Shrine of Our Lady of Fatima *or* Why Political
Questions Are Not All Economic," *Arizona Law Review* 23 (1981): 1283-1298;
and "Economic Theory and Environmental Law," *Michigan Law Review* 79
(1981): 1,393-1,419. Also see Steven Kelman, "Cost-Benefit Analysis: An Ethical
Critique," *Regulation* (1981): 33-40.

In general it seems possible to sort values into two categories—the preferences of individuals viewed as consumers and the ideals or commitments of individuals viewed as citizens.[23] Consumer preferences are expressed within a market, and the price one is willing to pay to satisfy them is a reasonably accurate indicator of their strength. But our decisions as consumers are constrained by the possibilities offered by the market. When driving down a highway lined with fast food restaurants, a person may choose to pay twenty cents more for a "hot and juicy" hamburger over the other fast-food options. But if, disgusted by the lack of options, the individual petitions for zoning laws limiting strip development and encouraging pedestrian malls in the hope that some ethnic restaurants and sidewalk cafés can survive, the value pursued is not so easily expressed in dollars and cents.

Since consumer preferences are expressed within the market and the market is determined by political, administrative, and legislative actions, it represents a serious distortion to consider values held by citizens to be expressed within the market: at least some of those citizen values are directed at controlling the market within which consumer preferences are expressed. They are transformative in nature. To treat them as preferences within the market is to confuse two quite different levels of value. For example, regulatory legislation governing the sale of pelts of an endangered species determines what preferences can be satisfied. The decision whether the pelts should be sold as fur coats (the question addressed by the citizen) is quite different from the decision whether to buy a fur coat once they are available in the market (the question addressed by the consumer). The former questions the assumption that consumer preferences are sovereign; the latter does not. That is, the former decision concerns the political question of what limits should be placed on the satisfaction of consumer preferences, while the latter concerns merely the satisfaction of preferences within these limits.

If all values are reduced to preference satisfactions, environ-

[23] Sagoff, "At the Shrine," p. 1,286. For a parallel but conceptually different account, see A. O. Hirschman, *Shifting Involvements: Private Interest and Public Action* (Princeton, N.J.: Princeton University Press, 1979), esp. pp. 62-76.

mentalists do face Ehrenfeld's dilemma: either species and eco-systems are without protection from sovereign consumer prefer-ences or they must be removed altogether from the market arena and given a value that is not instrumental in any way to prefer-ence satisfaction. Species preservationists need not despair, how-ever, provided they are willing to challenge the economists' re-duction of all values to preference satisfaction. And they should mount such a challenge because the transformative values that environmentalists attribute to human encounters with wild spe-cies and unspoiled ecosystems cannot be given proper interpre-tation according to the BCA approach. Besides the overwhelming technical problems involved in assigning dollar figures to ecosys-tem services and other demand values derived from wild species, the BCA approach is, in principle, incapable of expressing trans-formative values.

There are two reasons why transformative values cannot be ex-pressed on the unitary scale of values advocated by BC analysts. First, the preferences that constitute the unitary scale of values must be taken as given for the sake of the analysis. They must be fixed at a specified point in time, in order to be aggregated. Sec-ond, transformative values cannot be given expression in a mar-ket system because of its purely descriptive approach to values. The economists' preference-satisfaction model, which prides it-self on avoiding normative issues in policy analysis, treats all felt preferences as having equal claim to value.

For both these reasons transformative values cannot be ex-pressed within the BCA model. In order to use a single scale of value based upon felt preferences (demand values), analysts must take all values as being frozen at the point of computation and as having equal legitimacy. But transformative values make sense only if one believes that one set of preferences is better than an-other. To return to the example of the teenager who attends the symphony rather than the rock concert and finds her preference rankings altered, the value of the concert was to install a richer and more cosmopolitan value system.[24] But the value can be ex-pressed only if one is willing to assert that the broader set of pref-

[24] See Section 1.2.

erences (including rock music *and* classical music) is more valuable than the narrower one. Further, that value can be expressed only by comparing the teenager's preference system at different times. The value of the experience of attending the symphony concert is found in its transformation of the teenager's preference system at different times. The value of the experience of attending the symphony concert is found in its transformation of the teenager's preference system. At one time she had a narrower, less inclusive, and less defensible set of preferences, and at a later time she had a broader, more inclusive set that is judged to be more valuable.

Transformative values cannot be reduced to individual felt preferences because they represent second-level value judgments *about* such preferences and place limits on their free pursuit. If the judgments are expressed in regulatory legislation, they act as constraints on the market. While economists would say that regulation should only act to correct market failures (defined as failures to allocate resources efficiently), the view here defended implies that regulation can also constrain markets for reasons that defy expression on the unitary scale of demand values.

This broader conception of value is essential to environmentalism and, more particularly, to species preservationism. Environmentalists and species preservationists must argue that humans who encounter wild species and natural ecosystems undergo transformations of their value systems. And they must argue that, as a result of these transformations, the value systems of these humans are better. Because this transformative value cannot be expressed within the unitary, reductionist value system of demand values assumed by BC analysts, environmentalists must oppose that system. But they need not insist that other species have intrinsic value. It is sufficient to argue that they have transformative value for humans.

11.3 *The Preservationist's Case: Applications*

I have accepted as axiomatic that good policy results when rationally adopted objectives are pursued in consonance with the best available empirical data and theory. I have argued that according to species preservationists and environmentalists more generally rational objectives cannot be derived merely by computing individual felt preferences. Rather, a conscious effort to examine values in light of world views embodying metaphysical, epistemological, and scientific hypotheses is necessary. How, exactly, do experiences of wild species and natural ecosystems contribute to this process?

Humans are highly evolved animals. But, however important is consciousness, a characteristic that emerged from the incremental processes of evolution, it remains true that our values arise initially from our interaction with the environment. Just as intelligent pursuit of scientific knowledge regarding ecological, economic, and social systems requires knowledge of baseline data concerning how the systems function prior to intervention, intelligent debate concerning human values requires baseline information about the "natural" interaction of man and environment. Of course, in one sense such data are already lost. Human manipulation of the environment is pervasive. The situation is parallel to the realization that intelligent manipulation of the economy and social structure require useful data. Collection of true baseline data was not possible in the 1930s and 1940s, but comprehensive programs for keeping statistics from that point forward were instituted. Bureaucracies devoted entirely to the keeping of records were set up, and the earliest data gathered is now considered baseline data. While they cannot be claimed to give a picture of the social and economic system prior to intervention by government, these data function as a conventionally accepted reference point for comparisons with later statistics. Thus, while these data represent no absolute baseline, they do represent a relative baseline, a snapshot of how things are at a given point that can provide the basis for future comparisons and judgments about the effects of social policy.

Similarly, in the rational examination of human values it is important to maintain a basis for understanding the context in which values change through time. Sagoff established a crucial connection between experiences of nature and values held.[25]

[25] See Section 10.4.

Consequently, a rational examination of human values regarding nature requires an understanding of the different experiences of nature had by humans at different times, so that alterations can be recognized and interpreted. Preservation of ecosystems that have been altered to varying degrees by human intervention can provide snapshots, analogous to records kept by the Bureau of Labor Statistics, of the milieu in which human values have in the past been formed. Here true wilderness areas have a special claim because to the extent that they truly represent systems unaltered by management they can expose us to nature as our ancestors encountered it. Of course our experience of it will be different because experience is mediated by culture, but perceiving the context in which the values of past cultures (expressed in literature and art, for example) arose can be a powerful tool in understanding the interaction of nature and culture in value formation.[26]

Preservationists, therefore, value ecosystems affected by varied degrees of human intervention (especially those relatively free from it) because such ecosystems offer valuable experiences—experiences that have the potential to transform human values. By allowing individuals to encounter snapshots of nature from times past, natural areas provide the basis for both understanding and criticizing current values. Modern value systems that have developed in cities, in response to advertising blitzes and unlimited consumptive options, can be examined in a new light. What is it like to face, as a pioneer, a vast expanse of untamed wilderness? How does it feel not to be able to rely upon someone to tell you which product is the best (or the most socially acceptable)? Encountering wilderness, humans are forced to contrast their value system with other possibilities.

Encounters with species, likewise, provide occasions for us to reexamine our values. Wild species provide a vast storehouse of potential analogies and metaphors for human interactions with nature. In the same way that we might hope to avoid the mistakes of past cultures by studying history and finding analogies between human behavior in the past and human behavior in the present, we should find analogies in the adaptations of other spe-

---

[26] My discussion here was affected by a helpful discussion with Eugene Lewis.

cies. All species face the same ultimate problem: how to achieve high productivity within constraints posed by a limited and variable environment.

It is widely accepted that humans can learn how to accomplish certain tasks by imitating strategies of other species. For example, solutions to the problems of underwater communication are suggested by methods of the humpback whale, and the unbelievable strength achieved in spiders' webs has application to building processes. But these means-oriented opportunities to learn are only half the story. Every natural adaptation represents a balance achieved between needs and means to supply those needs. Needs are not unalterably given, and species can adapt to limited environments as much by modifying their structure of needs as by devising means to fulfill them. Boa constrictors, for example, can kill and swallow relatively large prey whole. This adaptation represents a clever means to obtain food, but the ability to go for long periods between feedings is an equal part of their adaptive success. Similarly, camels survive in the desert not by having developed unique means of obtaining water but by delaying their need for it.

To illustrate the point that humans can clarify their values by understanding adaptations of other species, I now present a detailed example. Biologists attribute to species very different reproductive strategies. Some categorize species as *r-selected* versus *K-selected*, while others argue that a tripartite classification is more accurate. While the evidence now appears to favor the latter system of categories, the simpler and more common distinction between r- and K-selection will serve for the purposes of this example.[27]

The concepts of r-selection and K-selection represent a continuum. Species near the r-selected end of this continuum have very

[27] See J. P. Grime, "Evidence for the Existence of Three Primary Strategies in Plants and Its Relevance to Ecological and Evolutionary Theory," *American Naturalist* 111 (1977): 1,169-1,194; and Geerat J. Vermeij, *Biogeography and Adaptation* (Cambridge, Mass.: Harvard University Press, 1978), p. 183. The type of reasoning involved in my illustration is not affected by the choice between a two-category theory and a three-category theory.

high rates of dispersibility. They tend to exist at the lower end of food chains. Individual members of these species live relatively short lives and invest less energy in building biomass. Instead, they expend large amounts of energy in reproduction, readily expanding their range into new areas, especially areas where disturbances have created open spaces. These r-selected species, such as crabgrass, disease-causing bacteria, and locusts, are often considered weeds and pests because they quickly take over and dominate favorable environments. These species usually have large populations that are subject to boom and bust cycles because they reproduce so quickly that they overshoot their food supply.[28] This also reflects their lack of feedback mechanisms to limit reproduction as the carrying capacity of their habitat is approached.

Contrasting with species at the r-selected end of the spectrum are K-selected species, which exist in smaller populations and have fewer offspring that live longer individual lives. Because they expend less energy in reproduction, they tend to build biomass and have larger relative size. They are also more specialized, involved in more mutualisms and coevolutionary arrangements with other species. Because of their greater biomass and higher degree of specialization, they are better competitors and, given time, they will force out r-selected species as, for example, trees force out the earliest colonizers after a disturbance as successional development proceeds from open space toward forests. K-selected species often have built-in mechanisms for limiting population growth. These limitations may be physiological (birth rates drop when food supply decreases) or competitive (large trees shade their own seedling and prevent them from reaching maturity).

This contrast provides a powerful metaphor for human population policy. Humans are, by their nature, K-selected. They have few offspring and live comparatively long lives. They are also

[28] See Thomas E. Lovejoy, "Species Leave the Ark One by One," in *The Preservation of Species*, ed. Bryan G. Norton (Princeton, N.J.: Princeton University Press, 1986). Lovejoy uses exactly this concept to discuss the tendency of human populations to grow beyond their life-support systems.

very specialized and exist high on the food pyramid. For most of human history, humans were highly adapted to particular, hospitable environments. There were many areas, such as deserts and the poles, where they could not survive in any numbers. But with the advent of modern technology, human population grew more quickly and expanded into marginal areas.[29] Population growth went into a boom cycle. And the inner, involuntary mechanisms controlling birth rates apparently have atrophied or become less effective as more of human functioning has come under conscious and/or cultural control.[30] Many population experts fear that the human species faces, in the near future, a population bust—a worldwide famine or war.

The application of this example is, I hope, obvious. Human birth rates are now controlled by conscious decisions. But ecological principles can inform those decisions. Limitations on birth rates in most K-selected species are involuntary, depending directly on signals from the environment as its carrying capacity is approached. Similar signals now abound for humans. Erosion occurs because marginal lands are brought into production, ocean fisheries are depleted, productive farmland is transferred to urban uses. These facts mirror the indicators that in other K-selected species would bring about drops in birth rates. But since human reproductive behavior can now be brought under conscious and voluntary control, humans have a choice to make. This choice, however, can be illuminated by parallels throughout the natural world. Will we emphasize the length and quality of individual lives, or will we pattern our behavior after r-selected species and continue the unprecedented growth made possible by our increased technological ability to alter the environment? Will we recognize that, as a K-selected species, we have heavy dependen-

[29] My suggestion that current human population policy is more appropriate to opportunistic species is anticipated by Charles Elton, *The Ecology of Invasions by Animals and Plants* (London: Methuen and Co., 1958), p. 144.

[30] There appear to be such physiological mechanisms still operative in contemporary hunter-gatherer societies where young are carried in close proximity to the mother's breast. This fact was pointed out to me by E. O. Wilson in informal conversation.

cies on other species lower on the food chains, or will we continue to reproduce and consume without concern for other species?

Given the broad approach to the endangered species problem that I advocate, an approach that sees particular endangered species as symptoms of the rapid human alteration of ecosystems, we can learn from analogies such as this one. If we continue to behave as if we are r-selected, the resulting upsets to the balance of nature will threaten more and more species. Eventually, these threats will affect us directly.

I am not claiming here that analogies such as this *decide* value questions. Humans, as rational beings, must submit alternatives to rational evaluation. But, since humans are evolved animals that are dependent upon a limited environment, it makes sense that the problems they face find parallels in other species' struggles to survive. And these struggles and the strategies evolved through them can illuminate human value questions by showing natural connections between choices and consequences. Other species provide patterns against which we can compare options. If unlimited populations have a natural tendency to overshoot the carrying capacity of their environment and if in nature this is correlated with shorter life spans for individuals, these facts at least provide relevant information for ethical discussion. In a sense, r-selected and K-selected species provide paradigms representing alternative life styles that humans can use as analogies to guide the conscious choices they must make.

11.4 A Coherent Rationale

I began both this chapter and this book by centering on a dichotomy between two types of reasons for preserving nature. Whether expressed as a distinction between economic and noneconomic, prudential and ethical, utilitarian and nonutilitarian, or resource and nonresource reasons, the intuitive dichotomy persists in one form or other. But in its usual forms the dichotomy encourages a false dilemma. Species preservationists appear to find themselves forced to choose between defending species merely as satisfiers of human demand values, in which case their support for preservationist policies is inadequate, or defending them as possessors of intrinsic value, in which case they face the nearly overwhelming task of clarifying unclear concepts and defending dubious principles.

A third kind of value—transformative value—can also be attached to other species, however. While demand values are, for the purposes of economic analysis, treated as fixed, they are in fact malleable and change through time. It is a central tenet (perhaps *the* central tenet) of the environmental movement that the values currently held in modern, industrialized nations are irrational and overly materialistic. Environmentalists also believe that a better understanding of the true human role in ecosystems would encourage belief in a more rational world view, one that clearly recognizes that the human species as it now exists is an evolutionary product of natural, environmental forces and is dependent on the survival of other species for its own survival. Encounters with wild species and natural ecosystems encourage acceptance of the ecological world view and cause humans to question the value of unlimited consumption. Further, humans can find in other species numerous analogies and metaphors that illuminate a wide range of value questions. Other species represent alternative modes of living within a limited environment. As fellow travelers in the odyssey of evolution, they provide guidance in understanding the human condition. Thus, while species fulfill demand values, they also help to transform those demand values and consequently have value on another scale.

The transformative value of other species and natural systems sheds light on a central controversy within the environmental movement. Some environmentalists, sometimes called conservationists, emphasize the value of natural products as resources for human consumption, insisting that they be used wisely so that they will provide continuous benefits for future generations. Conservationists seem to share the attitude of advocates of unlimited economic growth, differing only in their insistence on a longer-range accounting of costs and benefits. This attitude seems to differ radically from that of another group, sometimes called preservationists, who advocate setting aside ecosystems from human use. But now it is possible to see how these positions can be related in a coherent package.

Concerned that current human demand values be satisfied without destroying the productive potentials of nature, conservationists argue for efficiency and avoidance of waste in the process. They accept demand values and attempt to devise methods to fulfill those demands with minimal destruction to productive ecosystems. But their acceptance of these demand values is not absolute. They accept them for the purpose of analysis and planning only. As environmentalists, they also share with preservationists the belief that in an ideal world demand values would be less consumptive and materialistic. They can work to fulfill demand values efficiently in the short run, while hoping in the long run to effect changes in the preferences people in fact express. Hence, conservationists and preservationists pursue complementary, not incompatible, goals. Their differences in emphasis represent two aspects of protecting nature. There is no incompatibility between arguing for efficient use of natural resources in the fulfillment of demand values while also arguing that those values can and must be transformed.

Having discussed at length the various types of values served by wild species and natural places, it is now possible to address the question, What is the most coherent, complete, and adequate rationale for species preservation? Before answering it should be noted that this question is ambiguous. Are we asking for an idealized answer, the answer that would be given by a fully rational decider who has at hand all of the scientific information necessary to address policy questions, as well as a fully worked out metaphysics and theory of value? Or, are we asking what is the most coherent, complete, and adequate rationale at this time, given our imperfect scientific knowledge, our lack of a coherent metaphysics, and the lack of a clear consensus concerning natural values?

I choose to answer the second, more pragmatic question.[31] Hu-

---

[31] This choice should not be viewed as renouncing philosophy but rather as an affirmation that value questions affecting policy are often best answered within a concrete context. Value imperatives are no less philosophical for being applied to practical situations where knowledge and understanding are limited.

man demand values must provide an important part of any complete rationale for the protection of species. I have emphasized that all species should be assumed to have demand value because of their contributory value in supporting other species, some of which will prove useful. This demand value should not be underestimated. It cannot be denied, however, that demand values provide a shifting and ultimately unreliable basis for a policy of preserving species. Attributions of intrinsic value might well provide an adequate basis for such a policy—if they could somehow be divorced from their usual individualistic implications and invested with a clear meaning applicable to species. But the effectiveness of nonanthropocentric approaches would still depend upon the development of a set of generally persuasive arguments that rest on premises that can be made plausible to policy makers.[32] These tasks are unlikely to be accomplished quickly, for they demand what amounts to a full-blown attitudinal revolution. And yet active species preservationists emphasize the urgency of their task on the basis that a fourth or fifth of all species could be lost within the next twenty years.[33]

If I am correct in arguing that the species preservationists' case can be supported on less controversial, human grounds, then the unfortunate choice between inadequate demand arguments and adequate but unclarified nonanthropocentric arguments can be avoided. Environmentalists, who generally agree that current preferences are unfortunately consumptive and that wild species provide occasions for the reformation of these values, can insist that wild species be saved because they have transformative value. This strategy undermines the BCA requirement that all benefits be interpreted as quantitatively measured demand values and provides an interpretation of qualitative values derived from wild species. The argument for transformative values commits

[32] See Aldo Leopold, "Some Fundamentals of Conservation in the Southwest," *Environmental Ethics* 1 (1979): 132-141. Leopold discusses the difficulties in defending policies based on nonanthropocentrism to "men of affairs." Also see my "Conservation and Preservation: A Conceptual Rehabilitation," *Environmental Ethics* 8 (1986): 195-220, for a discussion of Leopold's decision to base his policy recommendations on the long-term good of the human species.

[33] Norman Myers, *The Sinking Ark* (Oxford: Pergamon Press, 1979), p. 5.

species preservationists only to the views that some value systems and preferences are better than others and that experiences of wild species have value because they play a positive role in transforming less acceptable into more acceptable values. Thus the proposed strategy appeals to consensually accepted values among environmentalists and incurs no new debits on their ledger of controversial claims.

This approach affords an adequate basis for the protection of species. Because all species have considerable contributory value—in supporting other species and natural systems—and because human moral and other nonconsumptive values are supported, in turn, by species and ecosystems, there exists a considerable obligation to preserve all species. This is a prima facie value: species should be saved provided the social costs are acceptable. Therefore, a full realization of the human values served by wild species and ecosystems implies a policy that will protect all species. It is unclear what, in addition, would be implied by the belief that species have intrinsic value.

Given this conclusion, little is gained by emphasizing the split between nonanthropocentrists and the inclusive version of anthropocentrism here outlined. The two positions would seem to have very similar policy implications. The question of whether a fully rational world view would include irreducible references to intrinsic values in nature can remain a matter for disinterested discussion and speculation. In the meantime, there exist adequate reasons for species protection based on human demand values and human transformative values.

Do references to the transformative value of nature require that nature be attributed intrinsic value?[34] It does not appear that they must. However, this question may only be answered conclusively after the full ontological and ethical implications of the ecological world view are worked out. References to transformative values require a belief that some value systems are objectively better than others. Environmentalists seem to share the view that con-

---

[34] As Tom Regan has argued. See "The Nature and Possibility of an Environmental Ethic," pp. 25-26, and my discussion in Section 11.1.

sumptive and materialistic value systems are inferior to less consumptive and less materialistic ones, other things being equal. If they believe this inferiority obtains because consumptive and materialistic value systems ignore intrinsic values in nature, the transformative value in question would depend upon intrinsic value in nature. But if they believe that consumptive or materialistic values are inferior for human, theological, or a variety of other reasons, transformative values would not require any support in nonanthropocentrism. For example, if environmentalists believed that humans live more satisfying lives if they are not bound by excessive greed for material things, this belief would provide an adequate, anthropocentric support for transformative values.

An analogy may clarify the role of transformative values in the overall rationale for preserving species. Suppose a family must move and fit all of their belongings in their station wagon. There is not room for both the television and the large, old family Bible. At first it is suggested that both should be taken to the pawn shop to see which is worth more. But another family member is aghast at the suggestion that the Bible, containing the family tree kept meticulously by generations of ancestors, should be treated as a mere commodity and closes an impassioned speech by claiming the Bible has intrinsic value and should be kept no matter what. Like this family, environmental philosophers often speak as if there are only these options: to make all decisions according to economic criteria or appeal to intrinsic values. There is an alternative: it can be argued that taking the Bible is the best thing *for the family*. If it is agreed that touching the Bible's soiled pages and seeing the great-great-grandfather's records of deaths in the family will give the children a sense of identity, of tradition and family values, the Bible should be chosen, regardless of short-term economic considerations. If it can be agreed that the Bible will have a beneficial effect on the family's values, it need not also be claimed that it has intrinsic value. It would then make sense to take the Bible with them and continue the discussion of whether it has intrinsic value en route. I have argued that an analogous situation now obtains regarding a policy of preserving species: we must act to preserve species now and can continue discussions about their intrinsic value while we do so.

The question may still arise whether transformative values are sufficiently clear, philosophically, to bear the weight environmentalists place upon them. One aspect of this question is whether transformative values may not rely implicitly upon intrinsic values in nature. Future discussion and debate will, hopefully, shed further light on the nature of transformative values, their relationship to intrinsic values, and their role in the arguments of environmentalists. In the meantime, I hope to have shown the importance of transformative values and that anthropocentrists and nonanthropocentrists can, given the current level of scientific knowledge and understanding of value questions, unite behind a strong and comprehensive policy of species preservation.

D

TRIAGE: THE PRIORITY ISSUE

TWELVE

FORMAL AND SUBSTANTIVE
PRIORITY SYSTEMS

12.1 *"Playing God"*

The magnitude of the problem of disappearing species, viewed worldwide, dwarfs resources currently available to address it. This grim fact ushers in the question of priority rankings in en dangered species policy, to the dismay of many environmentalists who often express considerable reluctance to discuss, far less advocate, priority rankings among species. Their discomfort at making and imposing "value judgments" on the natural order reflects a deep-seated uneasiness about human meddling in the natural world.

Much, but not all, of this discomfort originates in confusion. The purpose of this chapter is to eliminate unnecessary discomfort by clarifying the value commitments involved in various choices, while focusing attention on the areas where differences of opinion about priorities involve thorny issues.

Some of the discomfort results from a feeling that even to discuss the issue involves a form of hubris, an arrogant willingness to "play God." But most of this discomfort can be dispelled by distinguishing between individual and societal choices. Environmental activists and managers who face difficult priority decisions do not, individually, either create or control the governing conditions within which they must allocate the resources at their disposal. The collective choices made by members of modern society have created a situation in which perhaps over a million species will be extinguished in the lifespan of the present generation, and decision makers must act within that context.

243

Of course these individuals are participants in those collective decisions and share responsibility for the resulting condition with other members of society. It may be true that there rests a profound collective guilt upon the shoulders of mankind for creating this situation. Depending on the extent to which the individual, as a consumer and as a citizen, has contributed to the magnitude of the environmental crisis, he may be culpable.

But as an environmental activist or manager the individual does not create the situation in which he acts and, consequently, is not in that capacity responsible for having to make choices among species. He need not, then, accept blame for "playing God." To act in a situation with too many tasks and limited resources forces difficult decisions—but the responsibility for the creation of those conditions lies not with the actor but with humankind as a whole. The responsibility of the activist or manager in such situations is to make the decisions in the most rational manner possible.

This, the need to act rationally in making priority assignments, does involve hard choices and leads to an inescapable, practical discomfort. But even here clarification of the types of choices involved can relieve a considerable proportion of the anxiety that attends the responsibility for these decisions. My strategy will be to separate priority criteria into several categories and explain how some of these categories entail no controversial value commitments. To the extent that priority questions can be resolved on the basis of these criteria, the chooser is only acting sensibly on the basis of unquestioned values.

12.2 Formal vs. Substantive Criteria

Management documents use criteria that fall into four general categories: (1) considerations concerning the degree of endangerment and likelihood of success of recovery programs,[1] (2) taxo-

[1] The system described in Appendix I: "Priority System" of the *Program Management Document*, Endangered Species Office, U.S. Fish and Wildlife Service, June 1980, is based mainly on these considerations. For a full discussion, see William Ramsey, "Priorities in Species Preservation," *Environmental Affairs* 5 (1976): 601-602.

nomic considerations deriving from the comparative phylogenetic isolation of the species,[2] (3) considerations of ecological
value,[3] (4) considerations based on a range of cultural and socioeconomic values.[4] I will argue that priority decisions based on (1)
and (2) involve no controversial value judgments about species at
all. Only when priority systems are based on considerations listed
under (3) and (4) do controversial between-species value judgments arise.

In order to explain this claim it is necessary to distinguish *formal* from *substantive* criteria. I will say that a ranking of species
is *formal* if it can be accomplished without reference to characteristics of individual members of the species in question. For example, species can be ranked according to their numerical population level with no reference to the characteristics of their
members, but they cannot be ranked according to the average
longevity of members without referring to the lifespan of individuals. A ranking of species is *substantive* if it requires reference to,
and value judgments regarding, characteristics of individual
members of the species in question.

Priority rankings based on degree of endangerment involve
only formal considerations. The degree of endangerment of a species and the likely success of recovery programs designed for that
species can be measured independently of characteristics of the
individual members of that species. This is not to claim that such
assessments and predictions could be made if one knew nothing
about the fecundity rates, longevity, and so forth, of individual
members. Such scientific knowledge is crucial in ascertaining
both the degree of endangerment and the likely success of recovery programs. This is not the point. Rather, once such informa-

---

[2] These factors are discussed in a draft discussion, "Endangered Species Priority
System" (FWS/OES document number 301.3), June 8, 1976.

[3] Perhaps because of difficulties in reaching agreement on these scientific matters, few systems have given ecological value an important place. For a discussion
of this matter see ibid., pp. 9-11; and Bayard Webster, "Scientists Urge Triage for
Species Believed Endangered," *New York Times*, November 18, 1980, p. C1.

[4] For a discussion of criteria of this type, see "Endangered Species Priority System," pp. 9-11. Also see "Service Prepares Guidelines for Ranking Candidate Species," *Endangered Species Technical Bulletin*, U.S. Fish and Wildlife Service, Endangered Species Program, vol. 6, no. 8 (August 1981).

tion is gathered and a prediction is made, species can be ranked without further reference to those individual characteristics. For example, given knowledge about population size, fecundity rates, percentage of young expected to reach maturity, and so on, scientists can determine the probability that a species will survive for ten years if no protective or recovery program is undertaken. Likewise, given similar data plus possible recovery scenarios, approximate probabilities of success can be assigned to programs of certain predictable costs. Species can then be ranked according to the dollar cost per percentage point of increased likelihood of survival.

Both of these rankings, while based on scientific data, including data about individuals, are independent of that data in the sense that once the data are gathered and percentage figures assigned, there is an objective measure predicting the future course of a species under varied levels of human interference. Managers and activists can then judge where their efforts and dollars will have the greatest effect on species preservation. These judgments require no reference to attractive or unattractive features of members of those species.

Often, of course, the sort of data I have here mentioned will be largely unavailable at the beginning of the listing process. In such cases decisions will have to be made on less than complete knowledge. While this is unfortunate, it is not unusual and it does not change the basic point that the ranking in terms of degree of threat and likelihood of recovery is formal. Even when predictions are little better than "best informed guesses," they purport to be scientific, and activists and managers are justified in accepting the most plausible scientific judgments at hand, until better data become available.

Species can also be ranked, formally, according to taxonomic distinctiveness. A subspecies is less distinctive than a species with no subspecies. A species that is the last in its genus is more distinctive than a species that is one of several in a single genus. A species that is the last in its family is still more distinctive. Perhaps further refinements could be made according to whether the other taxonomic groups of that same level are also endangered. For example, if there are four species in a particular genus but all are endangered, each of those species might be given a higher distinctiveness ranking than an endangered species in a four-member genus where the other members are not endangered.[5]

Such rankings depend purely on the relative abundance of related taxonomic groups clustered around the species in question on the phylogenetic scale. They can be made formally. If the phylogenetic scale is visualized as a tree, the number of branches, that is, the number of taxonomic groupings close to a given species, can be determined without any knowledge of the characteristics of members of such species.

It is true that the decision to use phylogenetic diversity involves a value judgment—that diversity, in and of itself and wherever it occurs, is valuable. But there is also a sense in which this is a "formal value." It places no higher judgment on diversity in one taxon than another, provided each is equal in phylogenetic distance from related taxa. So, while priority rankings based on phylogenetic diversity rest on a value, it is an abstract value not appealing to the special nature of the individual members of the taxa involved. Persons who attribute value to diversity for aesthetic, biological, genetic, or economic reasons will accept the value underlying such rankings. This is not to say they would find such rankings definitive, as they may believe other values are more important than diversity. But I have argued in Chapter 3 that the value of maintaining diversity is considerable.

So far as difficult choices in the expenditure of limited energy and funds can be made using formal criteria, managers and activists need not shrink from decision making: they can point to objective rankings involving no substantive value judgments. While mistakes are of course still possible, it would be a worse mistake to ignore these objective bases for setting priorities, relying instead on whim or chance in allocating efforts at preservation.

Substantive value judgments are based on the comparative characteristics of individual members of the species involved. They rank species based on their ecological, socioeconomic, and

[5] "Endangered Species Priority System," pp. 8-9.

cultural values. Judgments such as these are likely to be much more controversial and subjective.

An example of a priority system that goes beyond formal criteria is the one that was in effect from 1981 until severe criticism during the 1982 reauthorization forced its abandonment.[6] Those guidelines, which were prepared "to assist in the identification of species that should receive priority review for listing under the Endangered Species Act of 1973, as amended," recognized two factors: degree of threat and "taxonomic status." While the latter criterion puts some weight on distinctiveness, it refers mainly to location on the phylogenetic scale. That is, while species are given higher priority than subspecies within the same category, the major categories are determined according to the "rank" of the species. Thus, priorities within each degree of threat are set as follows: (1) species of mammals, (2) subspecies of mammals, (3) species of birds, (4) subspecies of birds, and so on. This pattern is continued through fishes, reptiles, amphibians, vascular plants, insects, molluscs, other plants, and other invertebrates. The guidelines make clear: "Application of this priority system . . . would probably preclude listing activities related to species lower than category 11 (vascular plant species) during fiscal year 1982. Invertebrates and lower plants would not be listed nor would critical habitat be designated for previously listed species."[7]

Priority systems such as this can be justified only by arguing that the characteristics of members of some species make them more valuable, in some sense, than do those of other species. Concern is no longer directed at species merely because they are severely endangered or taxonomically distinctive. Under such a system, the activities of managers can no longer be justified by the general importance of saving species or of maintaining the most diverse gene pool possible. Species are treated unequally because of the value judgments humans make about them.

Assigning an "ecological value" to a species is, in theory, a sci-

---

[6] "Service Prepares Guidelines," p. 3. Also see U.S. Fish and Wildlife Service, "Endangered and Threatened Species Listing and Recovery Priority Guidelines," *Federal Register* 48, no. 184 (September 21, 1983): 43,102.

[7] Ibid.

entific matter or, more accurately, a cluster of scientific matters. The contribution a species makes to the biological productivity of a system is difficult to ascertain, but it is a scientific question. Matters become more complex, however, if there are, as is no doubt the case, several different ways in which a species may have ecological value. For example, plant species convert the sun's energy into usable biomass, while carnivorous predators do not.[8] But by exercising selective pressures on prey, carnivorous predators may contribute to biological diversity and specialization. Both of these are "ecological values." Value judgments arise in deciding which one should be promoted more vigorously because there is no uniform scale of ecological value. One suspects that, at least at this stage in the development of ecological theory, nonscientific value judgments affect which scale of ecological value is chosen. Priority criteria that rank species according to their ecological value must presuppose some indicator of that value, and such a presupposition embodies a substantive value choice. Since the choice among indicators cannot be resolved merely by reference to facts concerning various species' effects on ecological processes and since the choice indirectly implies that some species will be assigned greater ecological value than others according to characteristics of individual members of those species, substantive nonscientific and evaluative judgments can be hidden within it. Therefore, judgments of ecological value will be discussed together with other cultural and socioeconomic values.

12.3 Values and Priorities

Insofar as a ranking system simply assumes (a) that priorities in effort and expenditures should be sensitive to the cost and probable success of a recovery program and (b) that, other things being equal, as much phylogenetically defined diversity as possible should be protected, there will be little controversy. The first criterion implies cost-effectiveness in protecting the most possible taxonomic groups. The second qualifies the first, recognizing that

[8] Ramsey, "Priorities," p. 596.

all taxonomic groups do not contribute equally to diversity of the total gene pool of living things. Systems employing only these two criteria can be seen as anchored in the general value of diversity: other things being equal, a diverse gene pool is preferable to a nondiverse one. No decision as to *why* one values diversity is required.

But a ranking system that discriminates among species according to their ecological, cultural, and socioeconomic values offers certain advantages. It recognizes a widespread and perhaps important phenomenon: people do place different values on different species.[9] A ranking system based on such values would allow much finer distinctions to be made than a ranking system restricted to formal criteria. Such a system would allow us to make choices regarding expenditures among species of the same general cost factor and of the same genetic importance.

Before evaluating this proposal it is important to recognize that any attempt to make it operational implies that the values involved can be given reasonable, quantified estimates. That is, ranking species according to their ecological, economic, and cultural values presupposes that these diverse values can be placed on a single, cardinally quantifiable scale. Decisions based on formal categories of ease of recovery and degree of phylogenetic diversity merely reflect a common sensically corrected evaluation of how much a given unit of diversity is worth vis-à-vis other units of diversity. Formal priority systems do not require reduction of diverse values to a single scale, but such a reduction is inescapable in substantive priority systems.

This point is accepted by advocates of substantive rankings such as Gardner Brown, for example, who says:

> Everybody knows, including the General Accounting Office which documents some cases, that certain species are believed to be more important than others. Employees in the endangered species program select which species will and

---

[9] See Stephen R. Kellert, "Social and Perceptual Factors in the Preservation of Animal Species," in *The Preservation of Species*, ed. Bryan G. Norton (Princeton, N.J.: Princeton University Press, 1986).

will not be listed using personal criteria. . . . America's symbol, the bald eagle, simply ranks higher than the furbish lousewort in this nation's collective mind. Therefore, steps should be taken to acknowledge this truth by including a broader array of benefits in the analysis. It may be possible to give some but not all benefits a dollar expression. Perhaps the best which can be hoped for is a quality ranking of benefits or an index. Species might be given a benefit index value ranging between 1 to 100. Even if a working agreement can be reached about how to do this, a further step is required. *If only species with expected positive net benefits are to be preserved, the index must be translated into dollars.*[10]

If varied values, however they are assigned, are to be aggregated to create a single ranking scale, then they must be stated in commensurable units. Since costs are stated in dollar units, it is most reasonable and convenient to quantify benefits in like terms. Brown seems to recognize that some of these values can be assigned dollar equivalents only by arbitrary fiat. In these cases he would ask experts to rank species on a subjective scale. This subjective scale would then be translated (according to an undescribed formula) into dollar amounts. It is difficult to see why Brown prefers arbitrarily quantified figures to qualitative values.

Actually, there is one possibility that would avoid the need for a quantifiable scale. If it were assumed, first, that socioeconomic and cultural values vary together and, second, that they vary proportionately with placement on some other scale, quantification could be bypassed. This was the approach taken in the priority system in use at the Fish and Wildlife Service from 1981 to 1983.[11] There it was assumed that benefits derived from species coincide with placement on the phylogenetic scale, with relative proximity to humans on the phylogenetic scale representing greater value to humans. Thus, all mammals (species and subspe-

---

[10] Gardner Brown, Jr., "Survey of Methodologies for Ranking Species," Revised Report Prepared for the Department of Interior Office of Policy Analysis (May 1982), p. 13 (my emphasis).

[11] "Service Prepares Guidelines"; also see "Endangered and Threatened Species Listing."

cies) outrank nonmammals; birds outrank nonmammals and nonbirds, et cetera.

Unfortunately, both assumptions necessary for this ploy to work are false. Cultural and socioeconomic values do not vary consistently with each other or with location on the phylogenetic scale. Take, for example, aesthetic values as opposed to ecological/productive values. Ecological/productive values are concentrated at the end of the phylogenetic scale furthest from humans: plants are the only true producers.[12] But all the evidence suggests that humans place highest aesthetic value on species closest to them on the phylogenetic scale.[13] A similar point could no doubt be developed contrasting economic value with ethical value. Therefore, the ploy of using some other, unquantified scale has not worked and, given the diversity of values involved, seems highly unlikely to work.

I believe there are at least three convincing reasons why cultural and socioeconomic values of species cannot be reduced to a single scale. The first reason has already been developed in detail in Part A of this book and need only be summarized here. Given the prevalent lack of knowledge even among scientists as to ecosystem functioning and the role of particular species in that functioning, it is impossible to assign dollar values to economically productive benefits, let alone to more diffuse cultural values.

The second reason why cultural and socioeconomic values are not quantifiable can best be explained in reaction to Brown's explanation of how one might quantify them. Brown begins by noting that workers in the Office of Endangered Species, in response to an informal poll a few years ago, were willing to rank species according to their importance. He takes this to imply that experts do believe a ranking system is possible and useful.[14] Then he states that since the level of governmental appropriations is less

---

[12] See Peter Raven, "Prepared Testimony Presented to the House Committee on Merchant Marine and Fisheries, Subcommittee on Fisheries, Wildlife Conservation, and the Environment, Oversight Hearings on the Endangered Species Act," February 22, 1982.

[13] Kellert, "Social and Perceptual Factors."

[14] Brown, "Survey of Methodologies," pp. 11-12.

than would be necessary to save all threatened species, it contains an implicit recommendation that choices among species be made. He concludes that a triage strategy for species is necessary.[15] He then recounts remarks by experts concerning which species are most valuable in their eyes.[16]

The recourse to experts raises several important issues. First, it must be decided what fields of expertise should be represented in deciding triage issues. Second, someone must decide who counts as an expert and who doesn't. Brown seems to assume that the common person is ill equipped to make triage decisions. But it is necessary to ask *why* he is unqualified.

Is the common person unqualified because he prefers the wrong values? Or is it because, in spite of having acceptable preferences, he does not know which species contribute to the satisfaction of these preferences? By going directly to experts, Brown ignores or obscures this distinction and assumes that experts are best qualified both to designate values to be protected and to decide which species best protect those values. Problems arise on both levels.

Take first the problems of deciding which species serve given values. If experts are chosen from different fields of specialization, they are likely to embody the prejudices of their profession. If one asks a pharmacist which species contribute most to human health, one is likely to get a very different answer than from a nutritionist. Decisions about which species serve given values will be highly subjective depending on who is a "relevant" expert.

But common people's perceptions of what is valuable may be flawed not just because they are ignorant of scientific opinion on which species will serve which goals. They may instead be the result of a misguided value system. Economists, who espouse value neutrality, are loathe to criticize preferences, but it is clear that society in fact does so. Suppose a nonexpert ranks bears at the top of his priority list because his favorite sport is bearbaiting. If experts must confine their criticism of his preference ranking to cor-

15 Ibid., p. 16.
16 Ibid., p. 18.

recting the scientific facts, there would be no basis for criticizing this particular ranking, as bears clearly serve the goal of enjoyment of bearbaiting. But society has concluded that the pleasures derived from bearbaiting are cruel and are not to be countenanced as legitimate values. Likewise, a society might conclude that a recovery program for alligators should not be undertaken if the only goal it serves is to reestablish trade in alligator shoes. It might be judged that this purely consumptive luxury value is wrong, or at least should be ranked lower than other values.

These examples point up the most basic problem with Brown's suggested procedure. The use of scientific "experts" masks the fact that the original problem in ranking species according to their cultural and economic values was the incommensurability of the *values* involved. Should a species high in aesthetic value be ranked ahead of a species high in ecosystem productivity values, for example? An ecologist can render an expert opinion, perhaps, on which species is most likely to be productive in an ecosystem. And he can make a plea on behalf of the value we should place on ecosystem productivity. But how does this latter opinion tell against that of the aesthetician, who undoubtedly will favor attractive or interesting species? Brown's approach only reduces the problem of ranking species to the problem of which experts to choose and to the even more intractable problem of whether there are or should be "experts" concerning which values are most worth protecting.

The problem of choosing experts, then, suggests yet a third problem with substantive value systems. They fly in the face of the central concern of this book. It was argued in Part C that the value of species cannot be reduced to a matter of individual felt preferences because other species shape as well as fulfill preferences. Discussion and debate must concern *which* preferences are to be fulfilled as well as *how* they are to be fulfilled. If one ignores transformative values, recognizing only demand values (felt preferences), one is sure to undervalue species systematically. There are very difficult problems involved in estimating many of the values humans place on wild species. But they will not be solved by turning them over to scientific experts, who have no more basis for preferring aesthetic values over productive ones (or vice versa) than does the perhaps misguided nonexpert. Relying on scientific experts to rank the entire range of human values requires extending their responsibilities beyond the scope of their competence.

If values cannot be reduced to felt preferences, then they cannot be reduced to a single scale. If they cannot be reduced to a single scale, they cannot be given commensurable quantification. But advocates of a substantive priority system admit that such a system requires a quantifiable scale. Attempts to circumvent that requirement by using the phylogenetic scale failed miserably. This admission, however, dooms substantive priority systems, if the main arguments of this book are sound. In Part A, I argued that even demand values of species cannot be accurately quantified. In Part C, I argued that, even if demand values could be quantified, it would be a serious error to reduce the value of species to those values. Other species are also valuable for their role in shaping values. Here difficult philosophical, social, and cultural issues arise, and the conflicting values involved cannot be reduced to a calculus of preferences. Nor does there exist an identifiable class of experts who can produce nonarbitrary rankings. I conclude that there is no hope of constructing a defensible quantified scale for ranking species on substantive criteria. Substantive criteria for ranking species will not assist us in making difficult decisions concerning how funds and efforts to save species should be allocated.

Where, then, does the priorities issue stand? An assessment may suggest some new directions. First, it would appear that formal priority criteria will cause little controversy so far as they go. Formal criteria merely recognize and implement the general value of diversity. Any goal, including the one of preserving species, should be undertaken efficiently. Efforts should be expended on species that are in trouble but salvageable and on species that can be saved inexpensively. Within this context the suggestion that taxonomic categories more isolated on the phylogenetic scale be given higher priority can be considered little more than a gloss on "diversity." It is a decision that genetic diversity is more important than diversity defined in terms of the sheer number of species.

Even here, however, the pervasive problem of ranking values begins to obtrude. Reasonable individuals could differ in their value judgments when faced with the choice, for example, between saving several species that have near neighbors and one species that has none, if the costs are the same in each instance. This choice amounts to ranking the value of species diversity against the value of genetic variability. But, while reasonable individuals could differ, it is not unreasonable to hope that a goal can be agreed upon: the problem, ultimately, is only one of choosing a single concept of diversity to pursue.

Substantive criteria are more problematic for both theoretical and practical reasons. Theoretically, in applying such criteria felt preferences must be accepted as stated; otherwise values cannot be placed on a single scale. Since most environmental commentators believe that current preferences are unfortunately skewed toward the materialistic and consumptive, it seems impossible to support them as definitive of policy. Moreover, human values apparently favor, with little objective justification, species closest to us on the phylogenetic scale. It seems wrong to declare actual values indefensible on the one hand while continuing to advocate their use in hard choices about species preservation on the other. Appeal to experts can, perhaps, partially correct human biases when those biases are based on lack of information about which species fulfill specified values. But it is not clear how scientific experts can help if the problem is one of a skewed value system.

Practically, substantive criteria fare no better. First, they often tend to point in opposite directions, as was illustrated with the case of aesthetic and productivity values. Second, lack of scientific knowledge undermines attempts to place adequate value figures of any kind on species. One would require full knowledge of an ecosystem's functioning and the role of the target species in it before one could determine the value of the species in providing a single ecosystem service, for example. Even the proponents of such criteria admit such information is unavailable.[17] It is unclear

[17] Gardner Brown, Jr., and Jon Goldstein, "An Economic Argument for Preserving Some Endangered Species," unpublished manuscript (January 1983).

how they can advocate the use of substantive criteria while admitting that lack of knowledge make them impossible to apply.

To put the matter simply, the application of substantive priority criteria to decide which species to save requires a single, quantified scale of values. There are practical reasons for believing such a scale cannot be quantified and theoretical reasons for believing it is not unitary. The invocation of substantive criteria to determine which species to save, therefore, is simply not a live option.

But where does this leave the problem? It is admitted on all hands that current funding and staffing are insufficient to do all that is necessary to protect species from extinction. Neither are unlikely to increase dramatically in the near future. Under such conditions the decision to do anything is a decision not to do something else that desperately needs to be done. If formal criteria cannot decide all priority issues, it seems that environmental managers are doomed to making important decisions on whim and personal bias.

A stalemate seems to have been reached. A problem has been posed and the proposed solutions have been discussed and discarded for apparently good reasons. Before giving up in despair, however, it makes sense to look again at the formulation of the problem itself.

THIRTEEN

AVOIDING TRIAGE: AN ALTERNATIVE APPROACH TO THE PRIORITIES PROBLEM

13.1 Triage

Managers face excruciating decisions because appropriations for species preservation lag behind needs. The shortfall can be expected to widen as threats to species escalate in the face of human population growth and increasing consumption. Perhaps inevitably there have been calls for a system of triage, which takes its name from the French policy of sorting wartime casualties into three categories for medical treatment: those with superficial wounds that do not require immediate attention; those with wounds too serious to make treatment efficacious; and those in the middle range, having serious but treatable wounds.[1]

Once the issue is formulated in this manner, it seems obvious that effort is best concentrated in the third category. Scarce funds and energies should be targeted at saving those species that are both in need of saving and susceptible to being saved.[2] But the most arresting formulation of an issue is not always the most illuminating one; it will be useful to stand back from the triage formulation, which casts the problem of setting priorities as one of sorting species into categories, and ask whether there are other, more fruitful ways to look at the problem.

---

[1] See Bayard Webster, "Scientists Urge Triage for Species Believed Endangered," *New York Times*, November 18, 1980, p. C1.

[2] There is, of course, one important disanalogy. In medical triage the question never arises of sorting casualties based on value judgments concerning the comparative worth of the individuals involved (as was discussed in Chapter 12).

The endangered species problem (of which the triage problem is merely one manifestation) is not a single problem. It is more accurately seen as four closely interrelated problems: (1) what should be done when a species' population becomes so depleted as to threaten its continued existence; (2) what should be done to keep relatively healthy populations from declining and thereby falling into the threatened category; (3) how to avert, or at least slow, the predicted and potentially cataclysmic reduction of biological diversity over the next few decades; and (4) how to slow the trend toward conversion of natural systems to intense human use.[3]

In the triage formulation the priorities problem most naturally evokes question (1), because it considers threats to individual species. Once threatened, species require management initiatives designed to protect and nurture them individually. But the problem manifest in question (1) represents only one part of a larger trend toward ecological simplification. As one considers questions (2) through (4), one takes up a more holistic approach toward species preservation. A piecemeal, species-by-species approach seems appropriate to (1), although less so even to (2). It is not appropriate to (3) and (4). Correspondingly, the triage formulation, which sorts individual species, becomes less perspicuous as one considers the problem in holistic terms.

Once the problem of endangered species is looked at more broadly, as a problem of protecting overall biological diversity rather than as a problem of saving individual species, it can be stated in a different way: How ought insufficient funds and efforts best be spent to meet threats to biological diversity? This formulation does not point to a triage solution that sorts species into those that will receive attention and those that will not. The problem is how best to spend limited funds in the face of a general tendency toward ecosystem conversion and simplification.

Reformulating a problem, of course, does not make it disappear. Indeed, the problems faced under the new formulation are

[3] For a fuller discussion, see Bryan G. Norton, "Epilogue," in *The Preservation of Species*, ed. Bryan G. Norton (Princeton, N.J.: Princeton University Press, 1986). Also see Preface, above.

broader and more pervasive. But the new formulation has two positive effects. First, it focuses discussions of expenditures on the whole range of issues that can legitimately be considered problems of species preservation. The need to save individual species can be seen, more realistically, as one manifestation of a larger problem. Unless the underlying problems of habitat conversion and simplification are addressed, there will be progressively more species threatened with extinction, and it will become progressively more difficult to protect any given percentage of them. Second, the new formulation avoids the piecemeal approach to species preservation and its temptation to formulate the priorities issue as one of choosing among species. The real issue is how best to proceed, that is, how best to spend limited funds and to allocate limited efforts to alleviate the predictably serious loss in biological diversity. The present chapter addresses this more general question and poses tentative answers to it.

13.2 Expanding Horizons

There are several reasons why the goal of protecting biological diversity should not be reduced to the goal of protecting remnant populations of threatened species. First, if one thinks about the endangered species problem in this way, there is a tendency to treat it as merely a problem of protecting genetic diversity, with each species regarded as a repository for a set of genes. The problem of decreasing genetic diversity is, without doubt, a serious one. Growing demands for food production and attendant increases in monocultural agriculture exacerbate problems with pests. These and other forces make the protection of varied genetic resources in the form of new seed and breeding stock increasingly important. But biological diversity is a much broader concept than genetic diversity. Biological diversity is not just constituted by the number of species, subspecies, and populations extant; it is also constituted by the varied associations in which they exist. Viewed through time, a species existing in two different ecosystems can be seen as two different units of diversity. The selectional regimens and the attendant forces toward selective adaptation will be quite different. A species existing in an ecosystem represents not a static but a changing pool of adaptations. Treating a species merely as a static collection of genes, unaffected by environmental forces, is to ignore future adaptations. If a species exists in a variety of ecosystems and associations, it represents a whole series of different genetic dynamics and varied evolutionary trajectories.

Second, to treat the protection of biological diversity as no more than the protection of a few remnant populations is to ignore the whole range of contributory values species have when they exist over a broad geographical range and in varied associations. In chapters 2 through 4, I argued for the value of total diversity of areas. Having a broad range of species available in each area as potential colonizers and competitors for niche space strengthens the forces that lead, through niche packing, to diversity in successional communities. Each species has a contributory value as it comes in contact with other species: it offers a context of competition and opportunities for synergisms that create new adaptations and, eventually, new species. Species existing in varied habitats, then, are valuable because they give rise to long-term genetic variation. But diversity of biological life is also a valuable aesthetic and cultural resource. A diversity of plant associations, for example, provides visual and textural variety. If biological diversity is perceived only as the number of species extant, as a pool of genes to be utilized when needed, the whole range of aesthetic and cultural values dependent upon varied landscapes is ignored.

There is a third reason, related to the second, not to reduce the problem of protecting biological diversity to one of protecting gene pools. Managers often assume that concern for gene pools can be largely accommodated by preserving one or two healthy and stable populations of a species. If all of a species' natural habitat has been altered to a state where it can no longer survive unassisted, the species, seen merely as a gene pool, can be protected in zoos, preserves, and managed habitats. Indeed, some preservationists with a special interest in protecting gene pools speak as if the protection of species involves little more than preserving samples of seeds and germ plasm, reducing the value of a species to its basic genetic building blocks. But a species is more than this—it is a dynamic pattern of adaptations requiring a variety of habitats to realize its varied potentials. Protection of a few populations or a few germ plasm samples does not amount to protection of a species in the sense necessary for preserving biological diversity. Allowing new genetic adaptations is as important to long-term diversity as is protecting currently existing ones, and even genetic diversity is only inadequately protected by preservation in zoos and highly managed habitats.

Loss of genetic diversity is a manifestation of the deeper problem of decreasing biological diversity. As natural habitats are altered, converted, and simplified, many species suffer a decline in their number of independent populations. Attempting to protect genetic diversity through the protection of a few remnant populations ignores the most basic problem and will result only in a continual scramble to save individual species. A true solution would halt the tendency of more and more species to become so severely depleted that they require individual attention. If the deeper problems causing this tendency are not addressed, it can be expected that the effort to protect endangered species in remnant populations will become overwhelming.

Finally, the full transformative value of species will not be realized if the species is viewed singly, isolated from its habitat and adaptational interactions with other species. If humans are to learn from the adaptations represented in other species and from the analogies to human options stored therein, they must be able to observe these species in their natural habitat.[4]

For these reasons it is important to take a very broad look at the problem of declining biological diversity. By recognizing the forces that bring species to a threatened stage, a broader approach should keep more species from requiring individual attention. The triage formulation of the priorities issue would, in the process, be circumvented. Society would no longer face an inter-

[4] See Section 10.7. Also see Bryan Norton, "Learning from Nature: Zoos in a Technological Age," keynote address, Central Regional Conference of the American Association of Zoological Parks and Aquariums, Oklahoma City, March 4, 1985. Published in *AAZPA Regional Conference Proceedings*, 1985.

minable series of difficult choices among threatened species. Rather, the problem would be viewed holistically as one of halting the general tendency toward habitat destruction and loss of biological diversity.

Before exploring the effects of such a perspective on the priorities issue, it is appropriate to ask how the formulation of the problem relates to broader value issues in the society. Viewed in its most general terms, the problem of threatened and endangered species raises questions about the sort of relationship modern technological societies can and should have with nature. Will they see ecosystems as human habitats, as associations on which human life depends? Or will they see natural objects as no more than commodities available for use in the production of goods and services? If viewed in the former way, high value will be placed on the independence of ecosystems; they will be seen as having a holistic integrity that must be protected as the source of human life and as the basis of human values. If, on the other hand, productive potential is seen as the sum and substance of the relationship between man and natural objects, then nature is seen as a warehouse of consumable supplies.[5]

A warehouse is a place where resources not currently used may be stored, out of the way but ready at hand in case they are needed. Nature, so viewed, is a special sort of warehouse because humans assume they can go to it and find what they need, when they need it, drawing on its resources without fear of depletion. If the population of some species falls below a danger point, thereby threatening access to it as a commodity, then that species is "listed" for special concern; a fence is built around it, and it is isolated from further human contact. Nature is seen as a self-replen-

---

[5] These points are reminiscent of David Ehrenfeld's remarks on attitudes toward conservation. He uses, for example, a tool shed analogy to characterize the "humanistic" attitude toward nature (borrowed from Clarence Glacken). Further, he often expresses concern that nature is viewed merely as consumable goods and that the conservation effort is too often approached piecemeal. I agree with these points. However, I have criticized his alternative view, to partition nature into a resource and nonresource component (Section 11.1). See David Ehrenfeld, *The Arrogance of Humanism* (New York: Oxford University Press, 1981), esp. Chapter 5.

ishing supply of goods and services; if some individual element in that supply requires protection, it is provided by removing the species from normal interactions with humans. It is never asked why humans cannot normally cohabit with other species and the natural ecosystems that form their habitats; nor is it asked why more and more species suffer precipitous declines in population.

Once these attitudes are firmly entrenched, threatened species become nuisances. Necessary nuisances, of course (because they are building blocks—genetic material—for further technological advances), but nuisances nonetheless. So, they are isolated and confined, protected for future consumption. If they happen to be attractive to humans, they are placed in zoos where they can be watched and photographed—consumed aesthetically. But their removal from their natural habitat only accentuates and augments the underlying attitude that they are viewed not as elements of a functioning ecosystem but as discrete possibilities for future consumption.

These reactions can be traced to a very basic change in attitudes toward nature, which is no longer seen as the human habitat. It is no longer seen as a producer, sustainer, giver of life. Nature can produce, but humanly manipulated monocultures do it more efficiently. Nature can provide an endless variety of genetic resources, but these can be better protected in gene banks. Nature can provide aesthetic experiences, but it is easier to provide them in zoos. Protection of wild habitats takes space—space that can be used in technological forms of productivity.

Humans are not seen as one species, like others, inescapably dependent on natural systems. Highly developed technology increasingly insulates humans from the ways in which they depend on nature. Nature becomes not a place to live but a repository of raw materials to be extracted and used in technological forms of production. When it is recognized, in this context, that many species will be lost, the reaction is the same as if some nonliving commodity is threatened by destruction: some form of buffer is built between the threatened commodity and the forces that threaten it. In protecting species, rules against taking are instituted, and human activities jeopardizing particular species are relocated.

These reactions are appropriate, given that nature is viewed like other collections of commodities, as most safely protected in a warehouse, isolated from threat and ready for use when the need arises.

It would be beyond the scope of this book to attempt to criticize the world view embodied in the warehouse analogy. I assume that my attitude toward it is clear from my tone in writing about it, but I will not attempt a philosophical refutation of that world view. In contrast, I believe there are good reasons to explore alternative world views and, in particular, the sort of protection that would be afforded other species on the nature-as-human-habitat view. The reasons can be given quite practical formulation: the very recalcitrance of the triage formulation of the priorities issue indicates serious problems with attempting to integrate a policy of species protection into the nature-as-warehouse view. That species are becoming threatened at too fast a rate to allow listing and at too fast a rate to allow sufficient recovery programs indicates problems endemic to the triage formulation of the priorities issue.

It is not necessary to undertake a metaphysical critique of the nature-as-warehouse approach. The triage formulation of the priorities problem is, in a practical sense, a test case for that approach. I have argued that the triage formulation is intimately linked with the species-by-species approach to preservation. Now I am arguing that the species-by-species approach is a natural outgrowth of the world view on which nature is seen as a warehouse of raw materials to be protected because they may prove useful in the technological production of commodities or for aesthetic consumption. But the triage formulation fails. More and more species are threatened by habitat destruction caused by technological advances and by expanding human populations. Smaller and smaller proportions of species that require attention will receive it: the triage formulation leads to insoluble problems.

But a failure of one element of an interlocked system is a failure of the whole. The world view that sees nature as a warehouse of commodities is simply not, in the long run, compatible with the goal of species preservation. A society that desires to preserve biological diversity cannot do so by treating nature as a series of discrete commodities to be protected. Resources are not available to protect, on an individual basis, all of the species that will be threatened by a policy permitting wholesale conversion and alteration of natural systems. Providing for all of the needs of a species individually cannot be as efficient as protecting habitats where those needs are fulfilled naturally.

13.3 Habitat Protection

The species-by-species approach thus ultimately reduces to absurdity. But I have suggested an alternative formulation of the priorities question. Do not ask, which species should be saved? Ask, rather, how might agencies best spend the resources available to protect biological diversity? Habitat or ecosystem protection provides a more promising approach to preserving species than activities designed to protect individual species.

The advantages of a holistic ecosystem approach are numerous. Such an approach is more commensurate with a view of nature as human habitat. It recognizes that all species exist not in isolation but in functioning, interconnected habitats. Humans will continue to modify natural systems, but protection of large areas from human alteration signals a recognition that altered systems must exist in a larger context of naturally functioning, undisturbed areas. Such areas serve as reminders that human life grew out of and is sustained by the productive forces of nature.

Aesthetically, species should not be encountered only in zoos and botanical gardens, isolated from their natural habitats, but should be found in natural settings. Ideally, experiences of other species will not be treated merely as a consumable quantity—a trip to the zoo to fulfill preferences for experiences of predictably pleasing animals—but will have a sense of surprise.

Novel and unexpected encounters with species can jar the senses and the sensibilities and can begin the processes of value transformation and criticism articulated in Part C. However, once removed from its habitat and treated as a consumptive (or aesthetic) "commodity," a species does not provide occasions for value transformations. Unusual adaptations become merely aesthetic oddities. The pregnant analogies drawn from the struggle of species to survive in a limited environment will be missed. Species considered and treated as isolated from their habitat have less potential for transformative value.[6]

In general, species should not be seen as commodities held in waiting, ready at hand to provide goods and services as the need (human felt preferences) arises. Rather, they should be seen as having an independent existence drawing upon resources available in the natural ecosystems to which they also contribute. Humans would then be seen as existing within a natural community, altering some portions of it to increase productivity perhaps, but still existing in a context where nature is the ultimate producer and where all species, including humans, contribute to its functioning. This recognition encourages a view of nature as having an existence and purpose transcending human consumptive opportunities. Some theorists might interpret this purpose as intrinsic value, but this is not necessary. It can as well be thought of as providing the context in which humans consider and transform their currently felt preferences. In short, protection of natural systems, more than protection of isolated species, is consonant with the broadest range of human values: demand values, transformative values and, optionally, attributions of intrinsic value to nonhuman nature.

Above all, the habitat protection approach has a reasonable chance of success. Funds and efforts expended to protect species by protecting ecosystems and habitats are far more likely to be successful in the long run. In isolation from their habitat species require great amounts of care. Managers attempting to help isolated species must have knowledge and abilities often lacking in order to provide substitutes for the services provided naturally in undisturbed ecosystems.[7] The ecosystem approach protects species before they reach critical stages and require individual atten-

---

[6] See Norton, "Learning from Nature."

[7] See Lawrence B. Slobodkin, "On the Susceptibility of Different Species to Extinction: Elementary Instructions for Owners of a World," in *The Preservation of Species*.

tion. By addressing the problem in less acute stages, more efficiency per dollar spent can be expected. Efforts of this sort address not just the problem of how to save species once they have become severely endangered. They address all four forms of the endangered species problem simultaneously, by keeping healthy populations from undergoing declines, by protecting biological diversity generally, and by placing limits on how many natural systems are altered for human use.

Furthermore, funds and efforts expended to save habitats have many spin-off values. Saving a habitat is saving a collection of species, so that even species that are unknown and unidentified as endangered will be protected, along with better-known rare species. Ecosystem services will be protected, and, within the limits set by the principle concern, opportunities for transformative experiences will be promoted.[8]

It would appear, then, that when the question of priorities is posed as one of how best to expend funds and efforts, the answer is clear. They should be expended to protect as much and as varied types of natural systems as possible.

13.4 A Comprehensive Effort Outlined

A national effort is necessary to attack in a coordinated manner all four of the endangered species problems. The central offensive in such a campaign should be the protection of habitat. Domestically, this would require development of a set of categories identifying types of habitats and ecosystems and efforts to ensure that several systems of each type in each geographical locale receive protection. This book is not the place to go into a detailed analysis of methodologies, but the general shape of such a policy is relatively clear.[9]

---

[8] See Terry L. Leitzell, "Species Protection and Management Decisions in an Uncertain World," in ibid., for a more complete description of the practical values of the habitat protection approach.

[9] For a detailed discussion of the sorts of methodologies available, see Phillip M. Hoose, *Building an Ark: Tools for the Preservation of Natural Diversity through Land Protection* (Covelo, Calif.: Island Press, 1981).

Organizations like the Nature Conservancy and other private ecosystem protection groups have already done much to identify areas where more protection is necessary and have efforts under way to provide it. Federal and state governments should cooperate by offering financial and other assistance, by helping to coordinate ongoing efforts, and above all, by limiting the use of publicly owned lands that are appropriate for habitat protection. Public lands are a tremendous resource already available and could provide a considerable part of the solution. There will, of course, be some categories of habitats that require protection but that lie exclusively in private ownership. But a concerted effort to protect all types of ecosystems could involve a program of land trading. Private owners of lands requiring protection could be compensated with other, less sensitive, and perhaps more economically productive lands.

Within this general context, efforts to protect remnant species will retain some considerable place. It is necessary to have agencies designed to protect individual species because many species are already too threatened to survive merely through habitat protection. Further, it is likely that other species will, despite our best efforts to protect their habitats, become threatened in the future. At least some of these should be given protection.

The proposed plan does not ignore the importance of preserving individual species. My arguments, presented in Part A, that all species have prima facie value do not rule out discriminations among species. That all species have a value does not imply that all of their values are equal. When there are recognized reasons for treating a particular species as having special economic, cultural, or ecological value, there are special reasons to protect it, in addition to the general reasons applicable to all species. These special reasons may justify special protection and recovery programs.

They may also justify choices as to which habitats should be protected. If two areas representing the same habitat type differ only in that one contains an important endangered species, that is a reason to favor protection for the habitat containing the valued species. In essence this proposal reverses the current decision pro-

269

cedures in the Office of Endangered Species. At present species are identified and ecosystems are chosen and protected as critical habitat for those species. According to the proposed plan, habitats would be identified as deserving protection (because they are good examples of particular types of associations) and, when several areas are so identified, the existence of valued species can serve to determine which should be protected. Thus, the plan need not rule out special attention for individual species identified as having unusual value. These considerations may well affect which habitats are preserved and how they are preserved.

But it is a mistake to think of an office devoted to listing and protecting already endangered species as the core of the national program of species preservation. Efforts are better expended in protecting habitats, not individual species. The emphasis of the Office of Endangered Species would therefore shift considerably with the proposed plan. Less effort would be expended in listing species, and no assumption would be made that every endangered species should be given special protection. Indeed, I would suggest, though I admit it is a radical suggestion, that the listing process might be phased out. The Office of Endangered Species would then concentrate efforts on species that seem to have documentable value, for whatever reason. If a species is of known economic value, if it has some cultural importance, if it is pleasing aesthetically or distinctive genetically, then special efforts to save it are appropriate. But, unless there is some special, documentable reason to try to save some particular species, there should be no presumption that it is more valuable than some other, unidentified species existing in a threatened habitat. The efforts necessary to save species will be more effectively placed if they are directed at saving habitats because more species will, in the long run, be saved in this way. This may imply abandoning some species now identified as endangered and allowing events to take their course. But arguments that habitat protection is most effective are very powerful. More species will be saved by efforts directed at habitat protection than by efforts to identify, list, and develop recovery programs for every individual species.[10] Scaling down the listing

[10] See *Sunday Times* (London, U.K.), August 7, 1983, p. 9.

process would, presumably, free resources for an effort to coordinate federal, state, and local governmental and private efforts to protect habitats.

So far my practical remarks have been directed at domestic policy. Since far more species are threatened in other parts of the world, especially in the tropics, a complete endangered species protection policy must address the international problem. If the goal is to protect as much biological diversity as possible worldwide, then some efforts may have to be diverted from domestic programs to support a global effort. The problems of sovereignty and diplomacy involved are too complex to address here. Obviously, the U.S. government cannot act unilaterally within the boundaries of another nation. But international programs already exist, and many nations are highly cooperative on this matter. If supported by funds and efforts by the U.S. government, American leadership could make a tremendous difference in setting aside preserves of undisturbed habitat throughout the world.

It may be protested that the task set is too large, that it would cost too much in lost developmental opportunities. Perhaps this is the case. My purpose in this book has been to undertake a comprehensive and rational look at the goals and values supporting an endangered species policy. I have presented arguments that only a comprehensive effort to preserve habitats can succeed in the long run. It may well be that society lacks the commitment necessary to support such a comprehensive effort. In addition I have examined the reasons for saving species, discussed priorities in efforts at protecting endangered species, and suggested a program that will accomplish the legitimate goals of such a policy. I believe the arguments here presented support the sort of comprehensive policy I have outlined. I also believe that, compared to the benefits (considered in the broadest human terms over the longest run), the costs of a comprehensive policy to protect biological diversity may represent a remarkable bargain for the human species.

INDEX